I0120665

# IEG WORLD BANK | IFC | MIGA
INDEPENDENT EVALUATION GROUP

## The World Bank Group and the Global Food Crisis

### AN EVALUATION OF THE WORLD BANK GROUP RESPONSE

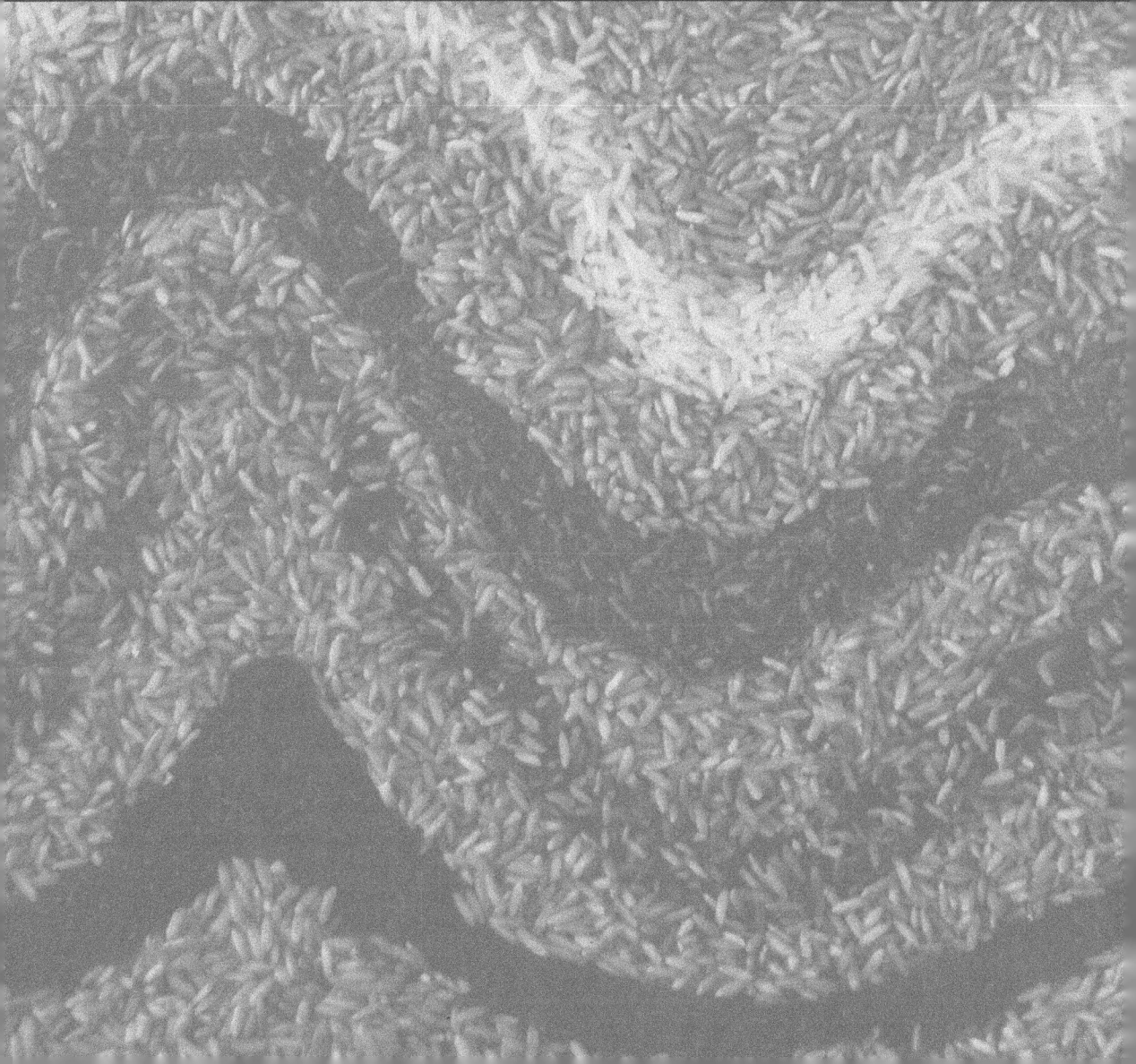

Design and cover photo: Crabtree + Company
www.crabtreecompany.com

**Library of Congress Cataloging-in-Publication Data**
The World Bank Group and the global food crisis:
an evaluation of the World Bank Group response.

pages cm
Includes bibliographical references and index.
ISBN 978-1-4648-0091-7 (alk. paper)
ISBN 978-1-4648-0092-4 (electronic)

1. World Bank—Developing countries. 2. Agricultural
development projects—Developing countries—
Evaluation. 3. Agricultural development projects—
Developing countries—Finance. 4. Food prices—
Developing countries. 5. Food supply—Developing
countries. 6. Economic assistance—Developing
countries. I. World Bank.

HD1431.W67 2013

363.8'526—dc23                    2013033682

# Contents

# Abbreviations

| | |
|---|---|
| AAA | analytic and advisory activities |
| ADB | Asian Development Bank |
| AfDB | African Development Bank |
| ARD | Agriculture and Rural Development (Department, World Bank) |
| ASAP | Agribusiness Strategic Action Plan |
| CAG | Agribusiness Department, IFC |
| CAS | country assistance strategy |
| CCT | conditional cash transfer |
| CDD | community-driven development |
| CFA | Comprehensive Framework Action |
| CGIAR | Consultative Group on International Agricultural Research |
| DFID | Department for International Development (U.K.) |
| DPL | development policy loan |
| DPO | development policy operations |
| EBRD | European Bank for Reconstruction and Development |
| ESW | economic and sector work |
| FAO | Food and Agriculture Organization of the United Nations |
| FSDPC | Financial Sector Development Policy Credit |

| | |
|---|---|
| GAFSP | Global Agricultural and Food Security Program |
| GDP | gross domestic product |
| GEM | Global Economic Monitor |
| GFI | Global Food Initiative |
| GFRP | Global Food Response Program |
| GFSI | Global Food Security Index |
| GIIF | Global Index Insurance Facility |
| GTF | Global Trust Fund |
| GTFP | Global Trade Finance Program |
| GTLP | Global Trade Liquidity Program |
| GWFP | Global Warehouse Financial Program |
| HDN | Human Development Network (World Bank network vice presidency) |
| HLTF | High-Level Task Force |
| IBRD | International Bank for Reconstruction and Development (of the World Bank Group) |
| ICR | implementation completion report |
| IDA | International Development Agency (of the World Bank Group) |
| IDB | Inter-American Development Bank |
| IEG | Independent Evaluation Group |
| IFAD | International Fund for Agricultural Development |

| | | | | |
|---|---|---|---|---|
| IFC | International Finance Corporation (of the World Bank Group) | | PWP | public works program |
| | | | ROC | regional operations committee |
| IFI | international financial institution | | RRRC | regional rapid response committee |
| IFPRI | International Food Policy Research Institute | | RSR | Rapid Social Response (Program) |
| IMF | International Monetary Fund | | SAF | Social Action Fund |
| ISR | implementation status report | | SIF | Social Investment Fund |
| LIC | low-income country | | SSN | social safety net |
| LTF | long-term finance | | STF | short-term finance |
| M&E | monitoring and evaluation | | TA | technical assistance |
| MAR | Management Action Record | | TASAF II | Tanzania Second Social Action Fund |
| MDB | multilateral development bank | | UN | United Nations |
| MIC | middle-income country | | UNCTAD | United Nations Conference on Trade and Development |
| NLTA | non-lending technical assistance | | | |
| OCHA | Office for the Coordination of Humanitarian Affairs (United Nations) | | UNDP | United Nations Development Programme |
| | | | UNEP | United Nations Environment Programme |
| OECD | Organisation for Economic Co-operation and Development | | UNHCR | United Nations High Commissioner for Refugees |
| PPAR | project performance assessment report | | | |
| PREM | Poverty Reduction and Economic Management Network (World Bank network vice presidency) | | UNICEF | United Nations Children's Fund |
| | | | WFP | World Food Programme |
| | | | WHO | World Health Organization |
| PRSC | Poverty Reduction Support Credit | | WTO | The World Trade Organization |
| PSNP | Productive Safety Net Program | | | |

# Acknowledgments

This evaluation is a product of the Independent Evaluation Group (IEG). The report was prepared by Ismail Arslan (Task Team Leader) with contributions from a team of evaluators and analysts. Management oversight was provided by Daniela Gressani and Monika Huppi during the approach paper phase, and thereafter by Martha Ainsworth and Ali Khadr. Aline Dukuze was responsible for all administrative aspects of the evaluation. William Hurlbut and Cheryl Toksoz provided editorial support.

Team members included Ana-Maria Arriagada, Marcelin Diagne, Kutlay Ebiri, Gershon Feder, Ade Freeman, Xue Li, Jennie Litvack, Xubei Luo, Ursula Martinez, Maximo Torero, Isabelle Tsakok, Joanne Salop, Jack Van Holst Pellekaan, and Melvin Vaz.

The 20 country case studies were conducted by Ismail Arslan (Philippines), Brett Libresco (Tanzania), Jennie Litvack (Kenya), Xubei Luo (Rwanda), Ursula Martinez (Honduras, Liberia and Nicaragua), Keith Oblitas (Sierra Leone) and Jack Van Holst Pellekaan (Tajikistan), Rahul Raturi (Bangladesh and Nepal), Hjalte Sederlof (Burundi, Djibouti, Lao People's Democratic Republic, Senegal, and the Republic of Yemen), Isabelle Tsakok (Ethiopia, Madagascar and Mozambique), and Utkir Umarov (Kyrgz Republic). Field visits were conducted to Kenya, Liberia, Madagascar, Nepal, Nicaragua, Philippines, Sierra Leone, Tajikistan, and Tanzania. A quantitative consolidation table based on country case studies' results was prepared by Ursula Martinez.

The World Bank's project portfolio and country case studies were identified by Ursula Martinez. Xue Li, Ursula Martinez and Melvin Vaz carried out the portfolio review for agriculture and social protection sectors. Saubhik Deb, Marcelin Diagne, and Xue Li conducted all quantitative analysis for the evaluation.

Peer reviewers were Alan Gelb (Center for Global Development and former Chief Economist for the Africa Region); Johan Swinnen (KU Leuven University and Center for Institutions and Economic Performance); and Peter Timmer (Center on Food Security and the Environment, Stanford University and retiree from Harvard University).

# Overview | HIGHLIGHTS

The unanticipated spike in international food prices in 2007–08 hit many developing countries hard. The International Bank for Reconstruction and Development and International Development Association of the World Bank Group organized rapidly for short-term support in the crisis, launching a fast-track program of loans and grants, the Global Food Crisis Response Program (GFRP). The GFRP mainly targeted low-income countries, and provided detailed policy advice to governments and its own staff on how to respond to the crisis. The Bank also scaled up lending for agriculture and social protection to support the building of medium-term resilience to future food price shocks. The International Finance Corporation (IFC) responded by sharply increasing access to liquidity for agribusinesses and agricultural traders in the short and medium term, as well as new programs to improve incentives for agricultural market participants. This evaluation assesses the effectiveness of the World Bank Group response in addressing the short-term impacts of the food price crisis and in enhancing the resilience of countries to future shocks.

Bank Group support for the short-term response reached vulnerable countries, though it is less clear whether it reached the most vulnerable people within countries. The program supported 35 countries, with Sub-Saharan Africa accounting for about 60 percent of the funding. The majority of support went to four countries—Bangladesh, Ethiopia, the Philippines, and Tanzania. The speed of the response often had costs for quality, and design deficiencies could not always be rectified quickly during implementation. The Bank's short-term assistance to agriculture took the form of input subsidy and distribution operations to increase food supply. Short-term support for social safety nets mainly consisted of in-kind transfers and public works programs. Existing public works and school feeding programs were continued or expanded, often in partnership with the World Food Programme (WFP). Only a few countries targeted support to infants and breastfeeding women—the most vulnerable segment of the populations. Most of this targeted support was for nutrition interventions.

The Bank's medium-term response for agriculture significantly increased lending and focused on expanding productive capacity and resilience. At the same time, analytic work declined, with adverse implications for policy dialogue and the quality of lending. The quality of the Bank's agriculture portfolio has declined, not only because of inadequate country analytical work but also because of resource and skill-mix constraints. In social protection, prospects for resilience are more promising, though risks remain, especially in low-income countries and for nutrition. Funding from the Rapid Social Response Initiative has enabled work on crisis-response capacity in low-income countries, which may help enhance future resilience.

IFC's short-term response focused on expanding agribusiness-related trade finance, working capital, and wholesale finance to increase liquidity in the agribusiness value chain with an increasing share in countries eligible for International Development Association (IDA) support. These programs had a high degree of additionality and received positive client feedback on the quality of processing and turnaround time.

Five lessons from this experience stand out. First, a detailed strategic framework for crisis response—which the Bank Group had in this case—is necessary but not sufficient for the effectiveness of interventions. Second, expansion in the scale of operations requires commensurate enhancement of administrative budgets to ensure success. Third, owing to the small amount of additional funding made available, many countries received only modest assistance that could not have had significant crisis-mitigating impact. Fourth, the effectiveness of increased lending depends critically on adequate analytical work and staff resources. Finally, for short-term responses to food price crises—as for other kinds of crises—having social safety net systems in place before a crisis hits is key to protecting vulnerable households and individuals.

These findings support four recommendations. First, ensure that country-driven food crisis response programs are sufficiently resourced with administrative budgets. Second, develop quality assurance procedures for food crisis response programs that mitigate the potential adverse effects of speedy preparation and implementation. Third, assist countries to better target the people most vulnerable to a food price crisis (especially children under two and pregnant and breastfeeding women) with appropriate nutrition interventions in their mitigation programs. Fourth, work with client countries and development partners to identify practical mechanisms (including indicators) for monitoring nutritional and welfare outcomes and impacts of food crises and mitigation programs, and work with them to implement those mechanisms and to report the results.

## Context and Background on the Food Price Crisis

International prices for food and other agricultural products increased by more than 100 percent between early 2007 and mid-2008. Prices for food cereals more than doubled; and those for rice doubled in the space of just a few months. Coming after a long period of low and fairly stable global food prices, the magnitude of the increases was unexpected, catching many governments off guard. The situation was aggravated by a concurrent rise in petroleum prices, affecting both consumers and businesses. Higher food prices might have served as an incentive to farmers, but this was offset by a spectacular rise in fertilizer prices, a key agricultural input.

The food price increases were particularly hard on the poor and near-poor in developing countries, many of whom spend a large share of their income on food and have limited means to cope with price shocks. An estimated 1.29 billion people in 2008 lived on less than $1.25 a day, equivalent to 22.4 percent of the developing world population. In addition, the Food and Agriculture Organization (FAO) estimated that 923 million people were undernourished in 2007. Simulation models suggested that poverty rose by 100–200 million people and the undernourished increased by 63 million in 2008. Although food and fuel prices leveled off in mid-2008, concerns about volatility remained; they surged again between June 2010 and June 2011.

In the last quarter of 2008, attention shifted to the global economic crisis and the ensuing recession. These developments broadened the scope of economic hardship to people in richer countries, while leaving the poor in low-income countries most vulnerable.

## The Global Response

During 2007, the FAO and forecasters within the World Bank and other agencies raised concerns about escalating food prices. But a concerted international response only began to take shape in 2008.

At the 2008 Spring Meetings of the World Bank and the International Monetary Fund (IMF), the Development Committee endorsed Bank management's proposal for "a New Deal for Global Food Policy, combining immediate assistance with medium- and long-term efforts to boost agricultural productivity in developing countries…" and urged donors to support the WFP to provide immediate assistance for countries most affected by high food prices. It also encouraged the World Bank Group to strengthen its engagement in agriculture.

Meanwhile, the United Nations (UN) Secretary General convened a High-Level Task Force and called for a Comprehensive Framework of Action. This marked the beginning of a sequence of international meetings, conferences, and working groups involving UN agencies, the Bank Group and the IMF, and parallel developments in G-7/8 and G-20 circles, focused initially on the construction of an action plan, subsequently on securing funding for it, and most recently on implementation.

## Evaluation Questions

This evaluation addresses three main questions:

- How did the Bank Group respond to the global food crisis?

- How effectively did the Bank Group help countries address the short-term effects of the food crisis?

- To what extent did Bank Group engagement during and after the crisis help countries to enhance their resilience to future food price shocks?

The evaluation analyzes the inputs, outputs, and intermediate outcomes associated with the Bank Group's response to the global food crisis, based on a review of the lending and nonlending portfolios, interviews with key stakeholders, and 20 country case studies. The assessment focuses on key aspects of the design, implementation, and early outcomes of the response. It distills from this experience lessons and recommendations for responding to future food price crises.

## The Bank Group Response

In May 2008, the Bank introduced as a central part of its response to the crisis, the GFRP set out a menu of fast-track interventions totaling up to $1.2 billion (including $200 million from the Food Price Crisis Response Trust Fund, financed from Bank net income). GFRP operations would be processed as "emergency projects," which have specific guidelines for preparation, appraisal, and approval. In April 2009, the Board increased the funding ceiling to $2 billion, which was available until June 2012. The Bank also called for an expansion of resilience-building agricultural and social protection coverage in its country programs under normal processing requirements.

The GFRP had three objectives:

- Reduce the negative impact of high and volatile food prices on the lives of the poor in a timely manner.

- Support governments in the design of sustainable policies that mitigate the adverse impacts of high and more volatile food prices on poverty, while minimizing the creation of long-term market distortions.

• Support broad-based growth in productivity and market participation in agriculture to ensure an adequate and sustainable food supply.

In pursuing these objectives, the GFRP supported operations in 35 countries. About one-third of the GFRP's 55 operations, the bulk of which were approved in fiscal 2008 and 2009, focused on food supply and pricing; one-third on social protection; and one-third on a mix of objectives. Of the 55 operations, 27 were freestanding and 28 were components added to ongoing operations.

IFC's response was mainly a sharp increase in agribusiness-related trade finance, working capital, and wholesale finance to increase liquidity in the agribusiness value chain and enhanced advisory services.

## Findings

▶ GFRP resources went to vulnerable countries, most of which received small amounts of support. Thirty percent of GFRP funds were allocated to the "most vulnerable" countries and a further 65 percent to "vulnerable" countries, based on a composite index of vulnerability developed by the Independent Evaluation Group (IEG) for this evaluation. About 60 percent went to Sub-Saharan Africa, the most affected region, where food expenditure accounts for over half of overall household spending; and about half of rice and 85 percent of wheat consumption is imported.

More than half of GFRP's resources went to four countries—Bangladesh, Ethiopia, the Philippines, and Tanzania. The remaining funds were distributed among 31 countries with large numbers of poor households facing serious food insecurity. Most countries accessing the GFRP received small amounts of assistance, generally less than $11 million per country. This was largely due to the limited availability of GFRP grant funds and the fact that in most IDA-eligible countries IDA resources were already largely committed to ongoing operations.

▶ GFRP operations were prepared and launched more rapidly than standard Bank operations. As emergency projects, GFRP operations were prepared using expedited processing rules. The median preparation time was 71 days, compared to 236 days for the Bank's broader portfolio. One-third of the evaluation's 20 country studies found evidence of trade-offs between the speed of preparation and the quality of project design and implementation.

▶ The range of cross-disciplinary skills needed to respond to the food crisis was stronger in the policy framework than in specific operations. The overall GFRP strategy and framework were commendably developed as a cross-sectoral and cross-network effort. However, in the design and implementation of specific operations, the range of sector and network skills was rarely available, with adverse implications for quality in some cases. The results frameworks underlying GFRP operations varied widely in quality and had particular design weaknesses in development policy operations (DPOs).

▶ There were particular shortcomings in the design and supervision of GFRP operations that took the form of additional and supplemental financing. In several cases, the "parent" operations were augmented without considering the implications of the additional activities for the results framework. This contrasted with freestanding GFRP projects, in most of which implementation status reports were regularly prepared, with evidence of due diligence by Bank management in reviewing them. More than half of the GFRP operations that were in the form of additional and supplemental financing were not mentioned in the implementation status or completion reports of the parent project.

▶ Many of the potential risks identified in the GFRP Framework Paper materialized. The Framework Paper highlighted potential risks to achieving results, such as limited availability of resources, capacity of client delivery structures, oversight arrangements, coordination among development partners, leakage in the targeting of beneficiaries, and inadequate component design. IEG's field-based project evaluations and country studies found that all of these risks were relevant, but many operations had been weak in addressing them. GFRP operations were granted the same flexibility with regard to Bank financial management and procurement policies as earlier emergency operations. This flexibility allowed the projects to shift the establishment and maintenance of financial management and procurement rules satisfactory to the Bank from the project preparation stage to the implementation stage. Indeed, almost all the pre-approval project documents that IEG reviewed stated the intent to have the necessary financial management expertise in place during implementation. However, IEG found no direct evidence that this commitment was kept in all GFRP operations.

▶ Fewer than half of GFRP operations have closed; two-thirds of those have been rated moderately satisfactory or better. Project outcome ratings reflect the relevance of operations' objectives and design, the extent to which they achieved their objectives, and (for investment lending) the efficiency with which they achieved their objectives. Among the 21 closed operations rated to date by IEG, two-thirds have been rated moderately satisfactory or higher on outcome. However, it is important to note that for operations with GFRP-financed components, these ratings reflect the performance of the entire project, not just the component. The outcome rating for GFRP operations are similar to project ratings for closed operations in Africa and in low-income countries in the Bank-wide portfolio. The quality of monitoring and evaluation for more than 60 percent of GFRP operations was rated by IEG as modest or negligible.

▶ The implementation of the short-term support program helped build experience for broader institutional crisis response mechanisms within the World Bank Group. In the past few years, the Bank Group has introduced several new instruments to mainstream some of the lessons learned from the GFRP. These include the IDA Crisis Response Window and the IDA Immediate Response Mechanism. These instruments have improved the Bank Group's crisis preparedness.

▶ The GFRP helped to reposition the Bank as a key player in agriculture and food security matters. The Bank Group's short-term response program in May 2008 was unique among global financial institutions in speedily articulating a comprehensive, concrete, and fast-disbursing financial support program to provide hard-hit clients with a menu of options for crisis mitigation. Along with the Bank Group's longer-term regular agricultural and social protection programs, and knowledge-based policy advice, the GFRP helped solidify the Bank's place as a key player in food security matters. The Bank's constructive participation in the UN High-Level Task Force and contribution to G-7 and G-20 meetings helped the international community to initiate several food security programs.

## The Response to the Short-Term Effects of the Food Price Crisis

▶ Pre-crisis analytic work by Bank staff provided a platform from which the Bank could offer timely policy advice. Food crisis mitigation policies were elaborated by the Bank as early as 2005. For example, the Agriculture and Rural Development Department issued a report in 2005 entitled *Managing Food Price Risks and Instability in an Environment of Market Liberalization*. This report anticipated that there would be occasions requiring short-term interventions, such as the use of publicly held strategic reserves and adjustments in variable tariffs. It warned that such short-term interventions should avoid undermining long-run market development. Prior analytical work on poverty and trade issues, notably by the Bank's Development Economics Vice-Presidency, was useful in assessing potential crisis impacts and formulating general mitigation policies and interventions.

With respect to safety nets, extensive analyses and lessons relating to the social impacts of and policy responses to previous economic crises indicated that, in the short term, the causes, transmission channels, and main poverty impacts of a crisis need to be assessed at the country level. They also indicated that the response needs to focus on protecting pro-poor social expenditures and on expanding large and effective safety net programs to operate in a "countercyclical" fashion as "automatic fiscal stabilizers." The studies also found that safety net programs, comprising cash transfers, public works programs, and human development interventions, needed to be country-specific.

Overall, a lack of data at the country level for assessing the welfare impacts of the crisis and hence for targeting specific interventions represented a significant constraint for the development of crisis responses in most GFRP countries.

▶ The policy advice provided through the GFRP framework document for the short-term response was pragmatic and cognizant of the need for second-best solutions. While export bans and price controls were considered undesirable under any circumstances, food subsidies were considerate acceptable for instance if targeted safety nets could not be expanded. Similarly, the use of strategic reserves to lower prices for all consumers was considered acceptable where better targeting was not possible. Input subsidies were recommended where credit and input markets were underdeveloped, given the long time required to establish functioning markets.

▶ The GFRP's short-term objective was to promote a supply response to alleviate crisis effects. Support for agricultural activities was granted to 27 of the 35 countries receiving GFRP funding. It was packaged into 32 GFRP agricultural operations, mostly investment operations of relatively small size (less than $6 million)—totaling $668 million. The very small size of Bank-supported operations limited their leverage and the operations' impact.

▶ Attempts to lower prices through tax and tariff reductions were not always effective. While the reduction of taxes and tariffs on food staples consumed mostly by the poor was recommended in Bank policy advice reports, it was also emphasized that these made sense in countries where the starting levels of taxes and tariffs were high. Nevertheless, the Bank supported operations in Burundi, Djibouti, and Sierra Leone where the rates were low to begin with, and where reductions did not affect prices but did reduce government revenues.

▶ Operations supporting agricultural supply response did not in most cases produce downward pressure on domestic food prices. The Bank's approach to support the distribution of agricultural inputs varied widely across countries. In some cases it supported the targeting of input subsidies to smaller and poorer farmers for redistributional reasons and in other cases to larger and better-off farmers for supply-response reasons. In some cases the necessary complementary inputs were not available, which precluded the full supply response from materializing. The low coverage of subsidy programs also likely limited their impact.

▶ Regarding social protection, the short-run objective of the GFRP was to ensure food access and minimize the nutritional and poverty impact of the crisis. Safety net activities were included in 60 percent of GFRP operations (33 projects in 27 countries). Of the 27 countries supported by the GFRP with social safety net activities, 23 were classified as either "most vulnerable" or "vulnerable" to the food price crisis, based on the composite index prepared for this evaluation. These two groups of countries received 96 percent of all GFRP funding for social safety net activities. Overall, GFRP social safety net funding increased Bank post-crisis financing for safety nets to low-income countries by 38 percent in the period fiscal 2009–11, compared to fiscal 2006–08.

▶ Country studies found that in most GFRP countries, analytical work to underpin social safety net lending was limited due to insufficient prior Bank engagement. The social protection interventions most frequently supported by the GFRP were in-kind transfers and public works programs, while cash transfers and direct nutritional support to young children and pregnant and breastfeeding women saw limited use. This mix of interventions reflects the dominance in the program of Sub-Saharan Africa, which accounted for more than half of GFRP operations with social safety net activities and almost a third of GFRP social safety net commitments. The Bank had limited previous engagement or analytic work in social protection in many of these countries, which constrained the choice of interventions and the ability to target vulnerable groups. Yet there was limited use of rapid assessments before launching these operations. Instead, the Bank used more general existing economic and sector work—in Bangladesh, Ethiopia, Kenya, Kyrgyz Republic, and Madagascar—or assessments by other donors—in Ethiopia, Kenya, Kyrgyz Republic, and Nepal.

▶ GFRP-supported in-kind transfers mainly involved the expansion of school feeding programs. The school feeding programs were often implemented in partnership with WFP. From a nutrition perspective, these programs do not ensure that the most vulnerable people—infants and pregnant women—are reached. From an education perspective, they may help raise enrollment and attendance, although they are not a substitute for a well-performing education program.

▶ Many safety net operations that aimed to target poor people in vulnerable countries relied on existing public works programs. GFRP projects also financed the continuation or expansion of existing food-for-work programs—through community-driven development and social investment funds in projects financed by the Bank, other donors (such as the WFP), or the government—designed to provide poor workers with an additional source of income even as they supported the creation, rehabilitation, and/or maintenance of public infrastructure. For public works programs to meet social safety net objectives, they need to have clear criteria for location, low wages to ensure self-selection of poorer workers, high labor intensity and use of unskilled labor, a portfolio of community-level investments, and sufficient duration to provide meaningful income transfers. These elements were not always present due to political economy considerations taking precedence.

In the emergency situation, these programs were constrained by lack of data for targeting. Most GFRP projects with social safety net activities used practical approaches to targeting such as combining geographic and community-based targeting, which led to risks that the intended beneficiaries would not be reached and/or that the non-poor might benefit. Weak monitoring and evaluation of interventions during implementation compounded this risk.

▶ Few GFRP social safety net operations targeted assistance to children under two and pregnant and breastfeeding women, who are the most vulnerable to malnutrition. Countries vulnerable to the food price crisis had the largest global malnutrition burden, yet only a few countries targeted nutrition support to children under the age of two and pregnant and breastfeeding women as part of their food crisis response. Only Kyrgyz Republic, Lao People's Democratic Republic (a pilot), Liberia (small sub-component), Moldova, Nepal, Senegal, Sierra Leone, and Tajikistan focused on infant and maternal nutrition.

▶ The key welfare outcomes related to the food crisis—poverty and malnutrition—were not sufficiently tracked to assess the welfare impact of the short-run response. Very few of the 20 case study countries provided an assessment of the impact of the food price crisis on the poor and vulnerable. Bangladesh, Nepal, and Nicaragua were exceptions. None of the 20 tracked malnutrition.

▶ IFC's short-term crisis assistance was mainly channeled through increased trade, working capital, and wholesale finance; and enhanced advisory services to agribusiness. Its liquidity financing operations supporting agribusiness and agricultural trade grew by 83 percent between fiscal 2008 and fiscal 2009. By fiscal 2010, its trade finance operation had grown by 160 percent relative to pre-crisis levels. While the effectiveness of trade finance operations specific to the agribusiness sector could not be assessed, IEG analysis of the main trade finance program (Global Trade Finance Program) suggested that it had a high degree of additionality and received positive client feedback on the quality of processing and turnaround time. Meanwhile, in line with its strategy supporting the building of medium-term resilience in the sector, IFC initiated programs expanding access to insurance against agricultural risks and focused its advisory services on high-productivity exporting countries. IFC's direct agribusiness investment strategy focused on the two ends of the food production spectrum—middle-income food-exporting countries able to affect global supplies and importing countries in Sub-Saharan Africa.

# Enhancing Resilience to Future Food Price Shocks

In the medium term, the Bank aimed to help countries build sound safety net programs and systems so that they would be better prepared for future crises. Earlier analytical work on agricultural development suggested several interventions and actions relevant to building resilience to food crises, including promoting agricultural productivity growth, and developing market-based risk management instruments such as futures markets.

▶ World Bank agricultural lending expanded significantly after the crisis and is now more directly focused on support to productive agriculture. Agriculture-oriented lending increased by 48 percent, from $8.8 billion in the pre-food crisis period of 2006–08 to $13 billion in the post-crisis period of 2009–11. The subsectoral composition of agricultural lending changed as well, and the share of lending directly supporting agricultural production increased. This trend suggests a potential increase in resilience.

▶ Deterioration in the quality of the expanded agriculture portfolio risks compromising the impact of Bank support on resilience to food crises. The volume of analytic and advisory activities in agriculture has declined in the Bank and is now focused more on nonlending technical assistance than on economic and sector work, with adverse implications for the knowledge base. In addition, the quality of Bank supervision of the agricultural portfolio has declined. The timing points to a crisis-related strain on resources available for supervision. These factors come on top of the recorded drops in performance for completed Bank agricultural projects—from 82 percent moderately satisfactory or higher in the pre-crisis period to 69 percent post-crisis. The main drivers of these trends (such as declines in the Bank's technical expertise and knowledge base) predate the crisis, but addressing them is key to enhancing resilience to future food crises.

While the quality of IFC's mature trade and liquidity finance operations related to agribusiness could not be specifically assessed, IEG's three-year rolling average of development outcome ratings for agribusiness investment projects shows no significant change between the pre- and post-crisis periods, with 71 percent of operations rated mostly satisfactory or better.

▶ Prospects for resilience of safety nets are more promising. Middle-income countries have continued to receive the largest share of social safety net lending post-crisis, but funding from the Rapid Social Response initiative enabled social safety net work on crisis response capacity in low-income countries, which may help enhance future resilience. Regular social safety net operations also show limited emphasis on improving nutrition among the most vulnerable—children under two and pregnant and breastfeeding women in the post-crisis period. The volume of analytical and advisory products increased considerably, but this was exclusively in the form of nonlending technical assistance. The social protection projects continued to perform well (76 percent rated moderately satisfactory or better on development outcome by IEG) relative to the Bank average of 70 percent in fiscal 2009-fiscal 2011.

## Country Focus and Partnership

▶ The Bank built on the ongoing aid effectiveness agenda in supporting country ownership and coordination with other aid donors. Against a backdrop of profound differences across countries in levels of development and in relations with development partners, the Bank built on the aid effectiveness agenda that has been progressing among donors, with a view to maximizing country ownership and minimizing strains on authorities' implementation capacity. In countries like Nicaragua and the Philippines, strong government oversight of donor activities shaped what the Bank and other donors did, ensuring (or not) coherence across partners' programs. Nevertheless, this approach sometimes led to frictions among donors, such as between the Bank and WFP over the government-determined geographic division of labor between them on school feeding programs. In other cases, existing donor coordination structures provided platforms on which the Bank's and others' response to the food and other crises could take shape. In the poorest countries, such as Liberia, Madagascar, and Nepal, there was considerable fragmentation across donors and donor programs—especially on safety net programs. According to partners interviewed by the evaluation team, in these situations the Bank played a constructive role, adding muscle to country authorities' efforts to establish greater coherence across donor-supported programs.

▶ In agriculture, coordination with other donors worked relatively well. At the level of the individual project or program, coordination was the norm for food and agricultural activities, especially with the Rome-based agencies. For the most part, coordination with the FAO and International Fund for Agricultural Development (IFAD) covered the provision of agricultural inputs—or in the case of WFP, school feeding programs—as the Bank and others provided only limited support for policy reform in the agriculture sector, given the very complex political economy of reform in the sector and country authorities' reluctance to tackle vested interests during the crisis.

▶ In social protection, coordination was more challenging. The partnership situation was different for social protection, for which there were far more donors and donor-supported programs seeking to help the poor and the vulnerable. In low-income countries, a common denominator was the school feeding programs pioneered by WFP and used by a number of UN agencies and bilateral donors—and by the Bank in Sierra Leone and other countries. The Bank approach in IDA-eligible countries also included food-for-work, social action funding, and support for the beginnings of social protection programs.

# Lessons

## Key Lessons

Clearly there will be other global food price crises in the future. What lessons can the evaluation offer about how the Bank Group should respond to them? Five stand out:

▶ First, a detailed strategic framework for the Bank Group's crisis response is necessary but not sufficient for the effectiveness of the interventions. This evaluation found that the GFRP Framework Paper was an important conceptual tool for organizing the Bank Group's response. However, there was often a disconnect between the intent of policy prescriptions in that paper and what was actually implemented, especially in short-term fast-tracked programs.

▶ Second, enhanced administrative resources—either incremental or redeployed from other purposes—and internal strengthening and collaboration are essential to an effective response that involves an expanded scale of operations. This lesson is reflected in evaluation findings for both the Bank and IFC. For the Bank, fast processing had a cost in terms of design quality, implementation, and results in some emergency operations. Moreover, launching such an ambitious crisis response program without a corresponding increase in the operational budget and staffing undermined the quality of existing lending and nonlending operations and had adverse effects on staff work-life balance. IFC's response benefited from the creation of a variety of trade finance facilities earlier in the decade; however, to some extent the benefits were limited initially by coordination problems across IFC units and between headquarters and regional offices. Subsequent consolidation of three investment departments and significant decentralization mitigated these issues.

▶ Third, limited additional resources and pre-crisis IDA allocations can constrain the ability of the Bank to respond to the crisis in IDA-eligible countries. Beyond the $200 million Food Price Crisis Trust Fund, the Bank Group did not secure additional funding to respond this crisis, and consequently adjustment in assistance to many countries was constrained by IDA allocations that had been determined by criteria unrelated to the crisis, and by limited flexibility within the ongoing country program. For most countries this resulted in modest operations that could not have a significant impact on food prices. This experience led to the establishment of Crisis Response Mechanisms that allow IDA countries access to resources beyond their standard IDA allocations.

▶ Fourth, the effectiveness of increased lending—as seen in the case of agriculture—depends critically on adequate analytical work and staffing. The crisis led to greater Bank Group emphasis on agricultural lending. But that emphasis was not supported by the increased staffing, analytic effort, and resources for portfolio management that were needed to ensure the quality and results of the new and ongoing operations in the sector.

▶ Fifth, in countries where social safety net systems are already in place, they can be critical to protecting vulnerable households and individuals during a crisis, but these are rarely in place in low-income countries and fragile states. As indicated in earlier IEG evaluations (on the Global Economic Crisis Response and on Social Safety Nets), the Bank provided major support for social protection programs in middle-income countries, matching growing country demand with innovative approaches and solutions. Although clearly established as a key priority for the new Social Protection and Labor Strategy, what clearly emerges from this evaluation is that the development of feasible approaches in the Bank's tool kit for use by low-income countries is work in progress. This remains a priority for the Bank's social protection team, with feasible interventions included in Country Assistance Strategies, thereby positioning countries to respond better to future shocks.

## Additional Lessons from the Bank Group's Short-Term Response

The findings also suggests early lessons that are specific to the short-term response.

▶ Senior management pressure to deliver particular crisis programs carries the risk of distorting program composition. The intensive promotion of the emergency program led to the inclusion of activities not addressing the crisis.

▶ Pre-existing country-owned agendas and ongoing programs can provide effective platforms for emergency operations. Building on a pre-existing country-owned agenda and the Bank's strong analytical work, the Philippines GFRP DPO achieved all of its short-term outcomes while catalyzing progress on the longer-term social protection agenda, including the establishment of an improved and expanded conditional cash transfer program.

▶ Context is important in considering the wisdom of tax and tariff reduction in a crisis response. A cautious approach is warranted, balancing likely pricing effects with possible implications for fiscal stress. In many cases, tariffs and taxes on staple foods were low to begin with, and rate reductions did little to help vulnerable groups, while aggravating the fiscal situation and threatening other government programs. Some emergency support compensated for budget shortfalls, but typically there was no a priori country-specific analysis to advise governments on the utility (and risks) of their tax and tariff policies.

▶ Good quality results frameworks and monitoring and evaluation arrangements for emergency operations are essential. The evaluation identified quality risks and concerns in results frameworks of GFRP operations (in both project lending and DPOs), especially where the crisis support took the form of additional and supplemental financing arrangements. The latter, often bore little substantive relationship to their "parent" operations, thus missing opportunities to identify emerging impacts (and problems) and the need for remedial action.

There were also problems in monitoring and evaluation, where, in several cases, monitoring surveys conducted after the closing of operations found evidence of sizeable leakages, as a number of beneficiaries targeted under the program and included in the distribution lists had not received food packages at the time when they were interviewed or had received incomplete packages.

▶ Simple, tried-and-true nutrition and health interventions are essential complements to social safety net programs in a food crisis and deserve wider use. The Bank's response to the food crisis had limited focus on targeting nutrition interventions, with Bank programs in only four low-income countries emphasizing nutrition support to children under age two and pregnant and breastfeeding women as part of their food crisis response program.

▶ Effective partnerships at the country level play a vital role in successful implementation of crisis-response programs. The donor coordination involved in Ethiopia, for example, was unique. In an effort to move to more predictable support and reduce fragmentation in humanitarian support, partners pooled their funds and came together in a unified stream of technical assistance supporting the government-led program. But partnerships were also important in countries where the authorities provided less leadership and the risk of donor fragmentation and duplication was greater—in these cases effective communications across donor groups and agencies is even more important for results.

## Recommendations

The findings point to four main recommendations to improve Bank Group effectiveness in responding to food crises.

▶ First, in cases where the Bank decides to respond to similar crises in the future: ensure that country driven food crisis mitigation programs are adequately resourced with administrative budgets to facilitate effective preparation and supervision of food crisis mitigation operations. The GFRP Framework Paper was an important conceptual tool for organizing the Bank Group's response, but implementation encountered problems. Operational resources were not expanded sufficiently for preparation and supervision to match the increased and accelerated volume of operations with adverse consequences for the quality of operations and staff work-life balance, and at the risk that other country priorities would be neglected.

▶ Second, develop quality assurance procedures for food crisis response programs that mitigate the potential adverse effects of speedy preparation and implementation. The Bank's fast processing of crisis response operations exacted a cost for design quality, implementation, and results in some emergency operations, suggesting that additional oversight of the standard quality assurance procedures was needed.

In some food crisis response operations, the Bank acquiesced with, or supported, policies and actions that were inconsistent with its own food crisis-related policy advice or that were not aligned with the country context. For example, in many countries, tariffs and taxes on staple foods were low to begin with and rate reductions did little to help vulnerable groups while aggravating the fiscal situation and threatening other government programs. In input subsidy operations, the underlying policy rationale was to stimulate a supply response to mitigate the adverse effects of input and food price increases, but the targeting was not consistently conducive to maximum supply response. The presence of other constraints (such as limited supply of quality seeds) was not always considered. Furthermore, the coverage of input subsidy operations was often too small to generate a significant supply response at the national level.

Where additional or supplemental finance instruments were used, the monitoring and evaluation arrangements, and the reporting on implementation and results did not consistently cover the food crisis response components of the project, limiting the potential for remedial steps and hindering impact assessment.

▶ Third, assist countries to better target the people most vulnerable to a food price crisis (especially children under two and pregnant and breastfeeding women) with appropriate nutrition interventions in their mitigation programs. Few Bank programs, in either low- or middle-income countries, emphasized nutritional support to children under age two and pregnant and breastfeeding women (the most vulnerable people) as part of their food crisis response program, even though most countries "vulnerable" to the food crisis have the highest global malnutrition burdens. Thus only a handful of low-income countries (Kyrgyz Republic, Lao People's Democratic Republic, Liberia, Moldova, Nepal, Sierra Leone, Senegal, and Tajikistan) focused on infant and maternal nutrition in their crisis response.

▶ Fourth, work with client countries and development partners to identify practical mechanisms (including indicators) for monitoring nutritional and welfare outcomes and impacts of food crises and mitigation programs, and work with them to implement those mechanisms and to report the results. The main welfare outcomes and impacts from the crisis—poverty and malnutrition—were not sufficiently tracked to assess the welfare impact of the short-run response.

# Management Response | EXECUTIVE SUMMARY

The Independent Evaluation Group's (IEG) evaluation focuses primarily on a temporary facility—the Global Food Crisis Response Program (GFRP)—created in May 2008 to help clients deal with spiking food and agricultural input prices and their negative impacts on food security. It also reviews the International Finance Corporation's (IFC) strategic food crisis response largely through the Global Food Initiative of mid-2008.

The evaluation does not examine the externally funded Bank-supervised GFRP projects, worth nearly $350 million and adding additional 14 countries to the response efforts.

GFRP handled more than $1.2 billion in Bank funding from the International Bank for Reconstruction and Development (IBRD), the International Development Association (IDA), and trust funds out of IBRD net income in Bank funding to 35 countries (60 percent Sub-Saharan Africa) over the FY2009–12 period, including grants out of the Bank's net income targeted to the poorest and most vulnerable countries. Externally funded Bank supervised GFRP projects brought the total number of supported countries to 49.

- GFRP projects reached 66 million people (49 million people benefited directly from agricultural interventions, and 17 million benefited from food-oriented social protection activities).

- GFRP funding from IDA, IBRD, and a trust fund out of IBRD net income accounted for 86 percent of the people reached. The remaining was reached by activities financed with mobilized external grant funds.

- The trust fund financed out of Bank net income allocated grants fairly equally (with due regard for size differences) among 27 smaller IDA countries. Other larger countries made use of the GFRP fast-track procedures to mobilize their own IDA allocations, or in some cases IBRD funds, for the crisis response.

• As stated in the IEG review, the median processing time for Bank-funded GFRP projects was 71 days compared to 236 days for all Bank projects over the same period, while IEG's outcome ratings for 21 closed GFRP projects evaluated were comparable to the overall Bank portfolio during the same time frame for the sectors in question.

The context for food crisis response has changed significantly since late 2007. Both the Bank and clients have made substantial progress in developing mainstreamed instruments and procedures to deal with high and more volatile food prices, after a period of increasing complacency (late 1970s to 2007) when international food prices were declining in real terms at a fairly steady pace.

There are now also mainstreamed instruments that did not exist in 2008 such as the IDA crisis response window, the IDA Immediate Response Mechanism, streamlined procedures for overall project restructuring, and enhanced flexibility in restructuring or cancellation of IDA operations with retention of resources in the country for on-going or additional operations. Similarly, the food price crisis served as a key catalyst for a broader strategic assessment of IFC's engagement in the agricultural sector that culminated in the 2011 Agribusiness Strategic Action Plan.

# Introduction

World Bank Group management thanks IEG for carrying out this ambitious and helpful evaluation of the World Bank Group Response to the Global Food Crisis. Management appreciates the extensive consultations at the concept note and draft stages. Management thanks IEG for the considerable effort expended towards close collaboration and the resulting exchange of views, which led to a more accurate, focused, and useful report of the GFRP.

The first section of this document sets out comments from management of the World Bank (the Bank). The second section provides IFC management comments. Management's specific response to IEG's recommendations is noted in the attached Management Action Record (MAR) matrix.

# World Bank Management Comments

GENERAL COMMENTS

▶ Scope. This report is clearly focused on the GFRP and less so on the regular portfolios of Bank financed operations in agriculture and social protection, or alternative trust funds such as the Rapid Social Response (RSR) trust fund. Furthermore, the evaluation did not include nearly $350 million in externally funded Bank-supervised GFRP projects.

▶ The 2007–11 Context. In 2007, countries had no cause to expect a change in international food and fuel markets. Food prices had been declining in real terms on trend and in a relatively stable fashion since 1980. Neither the clients nor the Bank in early 2008 had the chance to adjust their staffing, financial resources, or processing instruments to a rapidly changed set of external circumstances after major food, fuel, and fertilizer price spikes starting in the late fall of 2007. More than 60 countries experienced food riots in early 2008 associated with food price increases. Affected clients appealed desperately to Bank management for urgent help at the April 2008 Spring Meetings. However, neither clients nor the Bank were at that time ready to provide rapid response to the crisis in terms of instruments or resources, especially in the poorest countries.

Management acted rapidly, and by May 29, 2008, it presented for the Board's endorsement the GFRP Framework Document providing guidance to staff, particularly in crisis-affected field offices, on ways to respond to clients' urgent requests for assistance through budget support, social protection, and short term agricultural support. It also created a needed common understanding between management and the Board on how fast-disbursing response was to be achieved, and established a new suite of processing options tailor-made for the circumstance. The Board of Governors and Board of Executive Directors also approved a $200 million trust fund out of IBRD surplus to provide grants to the poorest and most vulnerable countries. As operations became effective, weekly reporting of new disbursements and concrete results were provided to the President and Managing Directors, decreasing to monthly reporting over the life of the facility. Executive Directors also received monthly reports over the same time period.

The GFRP allowed for a rapid response in recipient-executed projects for four reasons:

- The GFRP Framework Document provided contextualized policy advice and a menu of project design options for clients to choose from that helped Bank staff in the field accelerate the dialog with clients.

- It provided a set of procedures for fast-tracked processing, approval, and effectiveness.

- It allowed for very rapid communication between field teams and senior management, and with senior management relating regularly guidance and encouragement back to staff.

- It provided substantial additional Bank-sourced funding in the form of trust fund grants, and facilitated the use of IDA resources for restructured IDA projects.

The results of these actions were:

- GFRP projects and externally financed GFRP supervised projects in 49 countries reached 66 million people; of these 49 million people benefited directly from agricultural interventions intended to bolster food production, and 17 million benefited directly from food-oriented social protection activities.

- More than 95 percent of GFRP funding from IBRD, IDA, and trust funds out of IBRD net income (out of a total amount of $1.2 billion) was disbursed by closing of GFRP fast-track authority on June 30, 2012, financing 55 projects in 35 countries.

- As stated in the IEG review, the median processing time for Bank-funded GFRP projects was 71 days compared to 236 days for all Bank projects over the same period, while IEG's outcome ratings for 21 closed GFRP projects evaluated were comparable to the overall Bank portfolio during the same time frame for the sectors in question.

- An additional $350 million in external GFRP trust funds were mobilized, assisting 14 additional poor countries and financing additional GFRP operations in the other 35 countries financed from Bank-sourced GFRP funds.

- Increased visibility and effectiveness of the Bank in global discussions of how to deal with food crisis issues since 2008.

The GFRP represented both an innovation valid for a specific context and a carefully calculated set of risks. The GFRP Framework Document was an institutional innovation to create trust and understanding between the Bank's Board, staff, and clients when a rapid response was contingent on both the Board and some Bank departments adapting or accelerating regular procedures. It also bought time for all concerned to identify better approaches.

The specific temporal and institutional context dictated the menu of country options for investment that were allowed under the GFRP Framework Document, and the rules for their implementation. Management is pleased that the evaluation recognizes that context is key to the assessment of the GFRP, and that the GFRP Framework Document provided sound advice and a solid menu of options for the specific objectives it was designed to address within that context.

The present context is very different from that in 2008, as there is now increased awareness of food price uncertainty and volatility and their impact on the poorest. Both clients and the Bank have had time to adapt. There are now also mainstreamed instruments such as the IDA Crisis Response window and the IDA Immediate Response Mechanism that did not exist in 2008, the outcome in part of the GFRP experience.

The GFRP provided support to social safety nets (SSNs) which were missing in most countries before the crisis and helped (with other instruments, such as the RSR) to elevate the agenda of safety nets in international discussion and country level policy dialogue. It has proven to have lasting benefits, as 80 percent of countries in the world, according to IEG's own report on SSNs, are now considering measures to strengthen their safety nets.

While the GFRP Framework Document of 2008 is context specific and its applicability is unique to that period, it is clearly useful to the Bank and its clients to have an independent evaluation of its work and achievements.

While we broadly agree with IEG's findings, the evaluation does not take into account the less tangible aspects of GFRP implementation, which may have yielded additional knowledge and learning from this program. There are a number of additional issues that are relevant for the lessons to be derived by the implementation of the GFRP framework, including:

• Did the GFRP provide insights and time for the Bank itself to mainstream crisis response procedures better, such as with the IDA Crisis Response Window, the IDA Immediate Response Mechanism, its own internal approaches on food-related social protection, the Bank's reporting of food price movements, etc.?

• Did the GFRP help the Bank strengthen its cooperation with donor partners on food security? Did the Bank have impact on what other agencies did, such as the United Nations (UN) food agencies, through GFRP? Did it lead to new activities for the Bank?

• While the GFRP alleviated immediate pressures in 2008 on clients to take precipitous or unwise action with regard to the food price situation, did GFRP assistance help avert unfortunate country policy responses to soaring food prices of the type seen in 1974 (and that in some cases stayed for decades after) that began to resurface in 2008 (export bans, forcible procurement, food price fixing. etc.)?

• Did the GFRP improve trust about the fast response approach and the role of the Bank among Bank management and other key stakeholders such as the Board, clients, UN agencies, and nongovernmental organizations?

Overall, management feels that IEG's report provides useful information to answer positively about the first two issues. However, addressing the last two bullets would have required a different approach, including interviewing more extensively those who had been involved in using or approving GFRP.

The evaluation appropriately focuses on lessons learned about outcomes of projects in the crisis-specific context of the period, as opposed to a portfolio review of regular projects devoid of context. However, in addition to the interviews of UN partners, it could have addressed the less tangible aspects of what GFRP did for the Bank in terms of building trust and facilitating further collaboration.

## OTHER ISSUES

Management welcomes the recognition of the important role SSNs have played as an instrument to mitigate the negative effects on the poor and vulnerable. Management agrees with the acknowledgment that response to the crisis in most cases was in line with the broad strategic framework and at the same time pragmatic and adapted to country circumstances. Management also welcomes the recognition of the transformation in the Bank's social protection sector that was triggered by the crisis: increasing engagement with low-income countries (LICs) and the catalytic role of the RSR. Management recognizes the importance of building country-tailored systems in advance of the crisis and the need of long-term country engagement for the social protection agenda, as it takes time to build elements of systems and to secure political buy-in. The report also stresses a number of other important points regarding the SSN crisis response, such as limited availability of analytical work to inform the initial responses, incorporation of nutritional objectives in the SSN portfolio, and the importance of early warning information.

The evaluation's conclusion that the Bank failed to get additional resources for GFRP is erroneous, since the Bank did receive nearly $350 million in external trust funds for GFRP alone, on top of the additional funding for administratively separate but closely related food crisis response activities, such as the $60 million RSR.

In reference to a statement in the report that: "Four countries (Bangladesh, Ethiopia, the Philippines, and Tanzania) received more than half of GFRP's resources...," management would like to clarify that three of these four countries chose to request GFRP fast-tracking of projects funded by their relatively large existing country IDA envelopes, and the Philippines used GFRP procedures to accelerate an IBRD loan for food-oriented social protection. Additionally, the evaluation refers to "headroom" for fast-tracking existing country IDA resources as being a new "allocation" of funding. This imprecision of language matters where the evaluation implies that "GFRP allocated" a lot of financing to Bangladesh and Ethiopia, but less to Liberia and Sierra Leone. In fact, the grant funding to the latter two were grants fungible across countries and were truly "allocated" by the GFRP Steering Committee (composed of five Network Vice Presidents and a Managing Director), whereas the large IDA allocations of the first two was already at the disposition of the countries, which only had to follow GFRP rules to get the fast-tracking. The trust fund financed out of $200 million in Bank net income allocated grants fairly equally (with due regard for size differences) to 27 small IDA countries. Additional, external trust fund grants of nearly $350 million were allocated to 32 IDA countries, with some overlap with countries receiving Bank-funded GFRP.

The MAR focuses on four main issues:

1.  Adequacy and time alignment of operational resources.

2.  Bank quality assurance procedures for food crisis response programs.

3.  Protecting the most vulnerable groups.

4.  Monitoring nutritional and welfare outcomes.

Management is in general agreement with the specific recommendations in the MAR, although it wishes to highlight a disagreement with the statement of the evaluation findings in column one of the MAR under the second set of issues discussed. While management agrees with the recommendations for strengthening quality assurance as laid in the MAR, in column two, it does not agree with what could be read into the premise laid out in the "Findings" under column one, that "fast processing of crisis response operations exacted a cost in terms of design quality, implementation, and results".

In fact, the quantitative evidence in the report is summarized on the first page of chapter 2 of the evaluation: "The performance of two-thirds of the 21 closed GFRP operations was rated moderately satisfactory or better. These projects were prepared and became effective more quickly than the rest of the GFRP portfolio and most closed on time." This is substantially the same score result as the average for the much larger overall agricultural portfolio in the same period, which in turn showed a weakening of ratings at exit than the average for the three years before 2008.

Management concludes that while strengthening the overall quality and impact of operations remains a key priority for all sectors and for agriculture in particular, it is not clear that GFRP projects fared worse than other agricultural projects implemented since 2008. Furthermore, the evaluation does not provide conclusive evidence, other than anecdotally, that fast preparation was associated with lower impact at exit scores. Importantly, there is nothing in the evaluation to suggest that GFRP's specific "fast track procedures" themselves, at the heart of GFRP as an innovative approach, contributed to quality issues.[1]

The GFRP provided insights on how to better serve the Bank's stakeholders in an emerging world context that has become even riskier for the malnourished poor. While the policy advice and modalities of assistance in the GFRP Framework Document were appropriate (as recognized by IEG) for the context of 2008 and a couple of years afterwards, it is important to recognize going forward that that context has changed. Following the GFRP experience, the Bank now has a series of new mainstreamed instruments to help clients with response to crises, such as the IDA Crisis Response Window, the IDA Immediate Response Mechanism, streamlined Investment Lending procedures for project restructuring, and enhanced flexibility in restructuring or cancellation of IDA operations with retention of resources in the country for ongoing or additional operations. Like in the case of GFRP, future crisis responses will need to be designed to fit the specific context at hand. The trust developed between management and the Executive Board and the experience accumulated through the achievements of GFRP should provide the foundation to designing new responses, but the instruments will necessarily be different from what was laid out in the 2008 GFRP Framework Document.

## International Finance Corporation Management Comments

IFC management welcomes IEG's evaluation of the World Bank Group's Response to the Global Food Crisis. The report provides a useful independent assessment of IFC's immediate and subsequent activities in response to the unexpected rise in international food prices in 2007–08. The impacts of this crisis were especially difficult for the poor in developing countries, many of whom spend a large share of their incomes on food.

The report correctly recognizes IFC's strategic crisis-response through the Global Food Initiative (GFI) that IFC initiated in mid-2008. IFC expanded its agribusiness-related short-term finance (STF) in trade, working capital, and wholesale to increase liquidity in the food supply chain. It shifted its long-term finance (LTF) focus in agribusiness toward Sub-Saharan Africa and food exporting countries. It used its advisory services to build medium term resilience, help increase access to finance, enhance farm productivity, and improve the investment climate in the sector.

Management agrees with the report's overall positive findings on IFC's interventions. The report highlights the high degree of additionality and positive client feedback on IFC's STF response during this difficult period. It finds that despite the crisis, the development outcomes of IFC's LTF held up well based on IEG's data.

Management believes, however, that the report's focus on a fairly narrow time window means IEG may have inadvertently missed the most profound long term impact of IFC's institutional response. The food price crisis served as a key catalyst for a broader strategic assessment of IFC's engagement in the agricultural sector that culminated in the 2011 Agribusiness Strategic Action Plan (ASAP). Covering the FY2012–14 period, ASAP defines an integrated multi-sectoral approach to the sector to leverage development impact, support environmentally and socially sustainable outcomes, and increase food production and availability. It reflects a concerted effort to apply financial innovation and expertise across IFC departments. Underpinned by ASAP, the agriculture sector is now IFC's number one strategic priority.

ASAP focused IFC activities in the agriculture sector around three strategic priorities: i) enhancing food security, ii) increasing inclusiveness in the sector with greater benefit for smallholders and women, and iii) making environmental and social standards a business driver. These strategic priorities are addressed in an integrated approach, taking into consideration the different needs and capacities of countries. ASAP is consistent with the strategic themes of World Bank Group's Agriculture Action Plan FY2010–12.

## Endnote

[1] Management agrees with IEG that there has been an overall weakening trend in outcome ratings for the entire investment lending portfolio in recent years, as discussed in the 2012 IEG Report on Results and Performance of the World Bank Group. Management has been analyzing the root causes of this trend and has begun to address them as a matter of priority.

# Management Action Record

## Adequacy of Operational Resources.

### IEG FINDINGS AND CONCLUSIONS

The GFRP Framework Paper was an important conceptual tool for organizing the Bank Group's response, but implementation encountered problems. Operational resources were not expanded sufficiently for preparation and supervision to match the increased and accelerated volume of operations with adverse consequences for the quality of operations and staff work-life balance, and at the risk that other country priorities would be neglected.

### IEG RECOMMENDATIONS

In cases where the Bank decides to respond to similar crises in the future: ensure that country driven food crisis response programs are adequately resourced with administrative budgets to facilitate effective preparation and supervision of food crisis mitigation operations.

### ACCEPTANCE BY MANAGEMENT

WB: Agree

### MANAGEMENT RESPONSE

Management agrees with the notion that undertaking similar crisis response in the future should be matched with adequate resources, both financial and human, for project preparation and supervision. In the case of the GFRP implementation, the challenge was to mobilize internal resources commensurate with the administrative needs of GFRP project teams that were responding fast to the crisis.

The GFRP trust fund financed out of Bank net income funded 27 small emergency projects in the poorest and most vulnerable countries, 26 of them approved in the first seven months of GFRP before matching supplementary administrative funding became available from the President's budget.

Rapid adjustments of administrative budgets and staffing to respond to similar crises in the future will likely continue to pose a challenge under current resource allocation systems and a flat budget environment. Management is developing a single World Bank Group strategic framework and is progressively aligning business planning to it. This strategic alignment offers the opportunity to reflect and discuss with the Bank's shareholders how to address this challenge.

# Bank Quality Assurance Procedures for Food Crisis Response Programs.

The Bank's fast processing of crisis response operations exacted a cost in terms of design quality, implementation, and results in some emergency operations suggesting that additional oversight was needed over the standard quality assurance procedures.

In some food crisis response operations, the Bank acquiesced with, or supported, policies and actions that were inconsistent with its own food crisis-related policy advice or that were not aligned with the country context. For example, in many countries, tariffs and taxes on staple foods were low to begin with and rate reductions did little to help vulnerable groups while aggravating the fiscal situation and threatening other government programs. In input subsidy operations, the underlying policy rationale was to stimulate a supply response to mitigate the adverse effects of input and food price increases, but the targeting was not consistently conducive to maximum supply response. The presence of other constraints (such as limited supply of quality seeds) was not always taken into consideration. Furthermore, the coverage of input subsidy operations was often too small to generate a significant supply response at the national level.

Where additional or supplemental finance instruments were used, the monitoring and evaluation arrangements, and the reporting on implementation and results did not consistently cover the food crisis response components of the project, limiting the potential for remedial steps and hindering impact assessment.

## IEG RECOMMENDATIONS

Develop quality assurance procedures for food crisis response programs that mitigate the potential adverse effects of speedy preparation and implementation.

Specifically, the Bank needs to:

A.  strengthen ex-ante quality assurance oversight for food crisis response programs prepared under accelerated preparation procedures. Such oversight would ensure, inter alia, better alignment between the design of operations and the Bank's food crisis-related policy advice at times of spiking food, fuel, and fertilizer import prices, particularly with respect to taxes, tariffs, subsidies, and their targeting, considering the country contexts.

B.  ensure that food crisis response components, processed as restructured projects, additional or supplemental finance, include appropriate monitoring and evaluation arrangements; and

C.  require specific reporting on the crisis response components of restructured, additional or supplemental finance projects in implementation status reports, implementation completion reports and other project reports.

WB: Agree

While management agrees with the specific recommendations for strengthening quality assurance, it does not agree with the premise laid out in the "Findings" under column one that "fast processing of crisis response operations exacted a cost in terms of design quality, implementation, and results." In particular, management finds nothing in the evaluation to suggest that GFRP's specific "fast track procedures" per se, as opposed to rapid preparation, contributed to quality issues. Rather the answer must lie in other factors operating since 2008 common to GFRP and the regular portfolio.

A. Management agrees that the design of food crisis response programs prepared under accelerated preparation procedures needs to be aligned with Bank policy advice applicable to the country, temporal, and sectoral context. Management will consider how to optimize ex-ante quality assurance oversight for food crisis response programs prepared under accelerated procedures.

B. Management will strive to include specific monitoring and evaluation measures targeted to additional and supplemental funding that are appropriate for the project development objectives of that additional or supplementary funding, and that allow assessment of the separate contribution of that supplemental or additional financing.

C. Management recognizes the need to have separate results reporting for additional or supplemental financing, especially when different funding sources are involved.

## Protecting the Most Vulnerable Groups.

Few Bank programs, in either low- or middle-income countries, emphasized nutritional support to children under age two and pregnant and breastfeeding women (the most vulnerable people) as part of their food crisis response program, even though most countries "vulnerable" to the food crisis have the highest global malnutrition burdens. Thus only a handful of low-income countries (Kyrgyz Republic, Lao PDR, Liberia, Moldova, Nepal, Sierra Leone, Senegal, and Tajikistan) focused on infant and maternal nutrition in their crisis response. Likewise, only a few middle-income countries emphasized infant and maternal nutrition in their crisis response.

Assist countries to better target the people most vulnerable to a food price crisis (especially children under two and pregnant and breastfeeding women) with adequate nutrition interventions in their mitigation programs.

WB: Agree

Management will work with client countries to strengthen the targeting of nutrition programs supported by Bank projects responding to food price crises to the most nutritionally vulnerable populations (pregnant/lactating women and children up to 24 months) with a range of appropriate nutrition interventions.

## Monitoring Nutritional and Welfare Outcomes.
### IEG FINDINGS AND CONCLUSIONS

The main welfare outcomes from the crisis—poverty and malnutrition—were not sufficiently tracked to assess the welfare impact of the short-run response.

While theory and the Bank's policy guidance provide a framework to assess the impacts of food crisis on the welfare and nutritional status of key population groups, this requires country-specific assessments. Data scarcity is acute for most low-income countries. Thus, few GFRP countries assessed the impact of the food crisis on the poor. Some social safety net projects under the GFRP described mechanisms for the selection of beneficiaries, mostly using a combination of geographic and then community targeting, a practical approach that can produce serviceable targeting in data-constrained environments. However, the majority of projects did not specify actual and expected program "coverage" to assess the likely contribution of the project to the population in need of assistance. Most project documents state that project activities were targeted to food-insecure areas, but indicators only provide numbers of children to receive food in school or numbers of hospital patients to be fed.

### IEG RECOMMENDATIONS

Work with client countries and development partners to identify practical mechanisms (including indicators) for monitoring nutritional and welfare outcomes and impacts of food crises and mitigation programs, and work with them to implement those mechanisms and to report the results.

WB: Agree

## MANAGEMENT RESPONSE

Management agrees with the importance of tracking the impacts of food crises and of mitigation mechanisms on the welfare and nutritional status of key population groups. This will require country-specific assessments. Management will work with client countries and development partners to develop practical mechanisms for monitoring nutritional and welfare outcomes and impacts of food crisis mitigation programs. Specifically, Bank staff will work with client countries requesting assistance in handling food price crises to identify feasible indicators and design practical plans for data collection and analysis to implement the monitoring and reporting of the results of food price crisis mitigation programs.

# Chairperson's Summary:
# Committee on Development Effectiveness

On March 13, 2013, the Committee on Development Effectiveness (CODE) considered *The World Bank Group and the Global Food Crisis: An Evaluation of the World Bank Group Response* and *Draft Management Response*.

## Summary

The Committee thanked the Independent Evaluation Group (IEG) for the informative and insightful evaluation and welcomed management's response. Members appreciated the valuable lessons emerging from the evaluation, particularly with respect to issues around resource allocation, targeting of support, appropriate expectations of outcomes, and broader implications for project processing. Members congratulated management on the effectiveness of the Bank's swift response to the crisis. Members particularly welcomed the constructive cooperation between IEG and management during the process of finalizing the evaluation.

The Committee underscored the need for added flexibility in the redeployment of administrative resources—both financial and human—in crisis situations, given the Bank's constrained resources. While the report noted that the speed of the program may have come at the expense of quality, members felt that the Bank should not be afraid to admit that inevitably there are trade-offs with fast-track programs, particularly given unpredictable events or political economy concerns. Keeping in mind resource constraints, members urged management to balance the need for more active implementation and supervision with enhanced efforts on monitoring and evaluation.

Members warmly welcomed the reduced median project preparation time from roughly 236 days to 71 days for Global Food Crisis Response Program (GFRP) projects, and asked about the implications for non-crisis projects going forward. Members agreed that establishing country-level early warning systems would allow for rapid scaling-up of emergency responses in vulnerable countries, improved response capacity in crisis times, earlier planning, and more effective interventions. Members commented about lessons learned for the long-term and the need to focus more on improving agricultural productivity, infrastructure, and social protection programs. With respect to targeting the Bank's support, the Committee supported the call for improved in-country data collection and analysis, to better target and monitor outcomes.

Anna Brandt
CHAIRPERSON

# 1 Introduction

## CHAPTER HIGHLIGHTS

- The rapid rise of international food prices in 2007–08 was unexpected, catching many governments unprepared; it was particularly hard on the poor and near poor.

- Concern about the negative short- and long-term impacts of the crisis galvanized international action starting in 2008.

- The World Bank Group developed a framework for its response that was intended to rapidly provide funds to adversely affected countries.

- Special programs were created to address the short-term response; the Bank increased its lending in agriculture and social protection to support resilience to future shocks in the medium term.

- Design and implementation of the short-term response program helped to build experience for subsequent broader institutional crisis response mechanisms.

- This evaluation assesses the results to date of the Bank Group's short- and medium-term response to the food price crisis to inform the response to future crises.

# The Global Food Crisis

The dramatic increase in international food prices in 2007–08 was unexpected, coming after a long period of low and fairly stable global food prices. Prices for food cereals more than doubled between early 2007 and mid-2008; those for rice doubled over just a few months. In early 2008, the price of key agricultural inputs, particularly fertilizer, quadrupled and the price of fuel doubled. Food prices softened after June 2008, although they did not return to 2005 levels, but surged again between June 2010 and June 2011 (Figure 1.1).[1]

The global impacts were stark. Poverty rose sharply and is variously estimated to have risen by over 100 million people (Ivanic and Martin 2008) to some 200 million (Dessus and others 2008).[2] In addition, the crisis was reported to have pushed 63 million into undernourishment in 2008, a 6.8 percent increase over 2007 figures (Tiwari and Zaman 2010). Considering not only caloric intake but also food access and balanced nutrition, the negative impact on the poor was perceived to be even larger.

At the country level, however, the impact of the crisis depended on context. Countries that export a large share of their agricultural production benefit when prices are high, while countries that import a large share of food they consume suffer. In terms of fiscal effects,

FIGURE 1.1  Fertilizer, Food, and Oil Prices, 2004–12

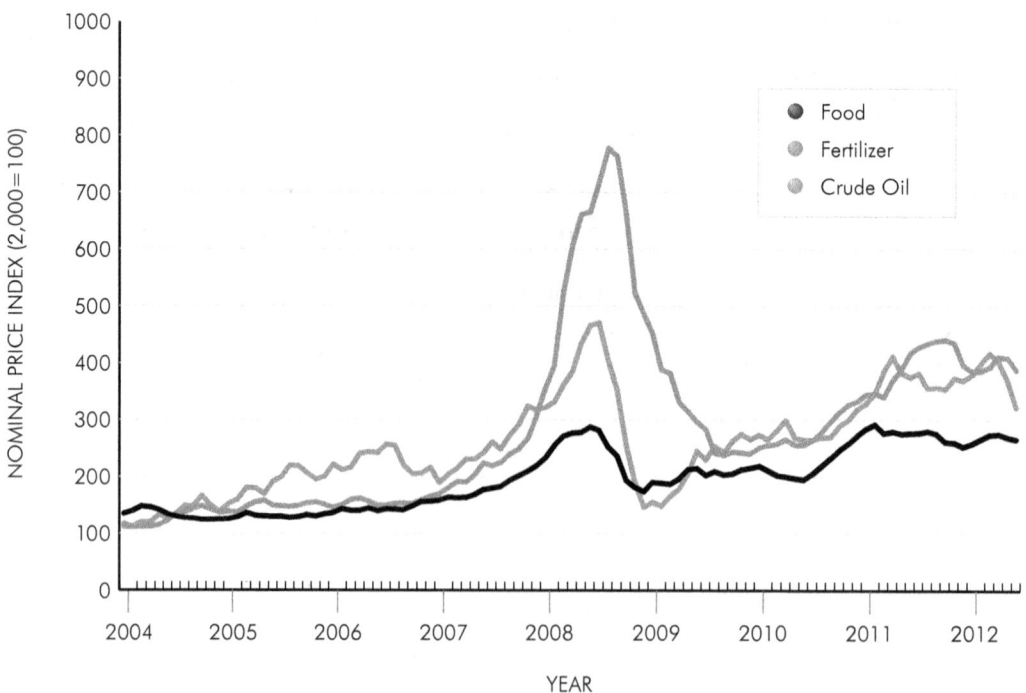

the impact of high food prices is strongest in countries where food subsidies are an important part of the budget. If the subsidies are increased to offset the higher price, the fiscal cost is even higher. For countries with limited fiscal space, this results in a reduction of funds for other needed investments, including agricultural research, education, health, and infrastructure, all of which can play an essential role in long-term growth.

Economists generally agree that the surge in food prices resulted from a combination of factors: higher fertilizer and fuel prices, diversion of agricultural land for feedstock crops driven by biofuel production, reduction of grain stocks in some Organization for Economic Cooperation and Development (OECD) countries and China, adverse weather conditions in some countries, re-emergence of grain diseases such as wheat rust in major producing countries, and stagnation in investments to increase grain productivity in developing countries (Appendix A and World Bank 2012). Once food prices increased significantly, further pressure on international prices arose from export bans and similar trade-curtailing policies by several major producers attempting to maintain lower domestic food prices.

## The International Response

Many governments were unprepared for the economic and political implications of the sharp food price increases of 2007–08, and many developing countries experienced domestic turmoil. Stakeholders in international development were concerned about its immediate adverse impact on the poor and its longer-term negative implications for human and economic development. In addition to the long-established Rome-based Committee on Food Security,[3] a United Nations High-Level Task Force, of which the World Bank was a member, was established in April 2008 to coordinate the response to the crisis. Bilateral and multilateral donors pledged substantial amounts to fund a variety of short-term interventions to mitigate the impact of the crisis as well as medium-term and longer-term interventions to increase agricultural production to help avert future crises.

## World Bank Group Response

To address the short-run impacts of the crisis, the World Bank launched the Global Food Crisis Response Program (GFRP) in May 2008, and the International Finance Corporation (IFC) launched its Global Food Initiative (GFI) shortly thereafter. The GFRP began to provide assistance to mitigate the effects of food price spikes six weeks after the Development Committee presented a "New Deal for Global Food Policy" at the Spring Meetings in April 2008. The framework document for the GFRP established a streamlined decision-making process and was a key tool, particularly in the design phase, reflecting a collaborative effort of many Bank departments.

In addition to the specific activities to address directly the short-run impacts of the food price crisis of 2008, the World Bank boosted lending in agriculture and social protection to build resilience to future shocks. The volume of agricultural lending (other than through GFRP) increased from an average of $3 billion per year in fiscal years 2006–08 to $4.3 billion in fiscal 2009–11. Social protection lending (other than through GFRP) over the same period increased from an average of $400 million per year in fiscal 2006–08 to $3.3 billion in fiscal 2009–11.

IFC focused its short-term response to the food crisis on expanding credit for agribusiness working capital and financing to facilitate trade transactions related to agribusiness along the whole supply chain. In fiscal 2009, trade finance operations totaled $758 million—an increase of 83 percent over fiscal 2008. This trend accelerated in fiscal 2009–11, with trade finance operations totaling close to $3 billion, compared to $596 million in fiscal 2006–08. In terms of building future resilience, in fiscal 2010 IFC provided nearly $2 billion in financing across the agricultural supply chain, including financing for projects to improve storage and distribution of agricultural produce, expand rural and agricultural trade finance, and expand food processing.

Between 2008 and 2011, the Bank Group set up a total of five special programs to implement its crisis response activities (Box 1.1), including GFRP and GFI. In addition, there were three externally funded trust funds providing $344.5 million funding to food-crisis-related activities in 32 different countries financed by Australia, Canada, the European Union, Korea, Russia and Spain. Approval of these externally funded projects began only in mid to late 2009, and most were approved well into 2010 and beyond. Apart from the GFRP and GFI, the operations financed by these externally funded trust funds are relatively new and little information is available on them, so they are not evaluated in this report.

In addition to its lending support, the Bank also produced reports analyzing the global, regional, and national causes and implications of the crisis and provided international stakeholders and national governments with advice on short-term mitigation and longer-term policy and strategies to avert future problems. Bank support to the Consultative Group on International Agricultural Research the Bank (CGIAR) helped to initiate new programs to address issues of food security and agricultural development. A timeline of the response of the World Bank and other donors is in Appendix B.

**BOX 1.1** Bank Group Food Crisis Response Programs

**Global Food Crisis Response Program:** The GFRP was launched in May 2008 with initial funding of $1.2 billion for three years. The short-term program was subsequently scaled up to $2 billion and extended through June 2012. The program's activities are a mix of technical assistance, development policy, and investment operations under four components. As of the end of fiscal 2012, the Bank had approved and funded $1.24 billion in GFRP projects.

**Global Food Initiative:** IFC initiated this program in mid-2008. This initiative comprised an umbrella for a range of activities and programs that were to constitute IFC's short-term response to the crisis. The initiative entailed: (i) an expansion of trade finance for transactions undertaken by agribusiness operators to help improve agriculture related trade flows; (ii) expansion of advisory services to agribusiness to increase the local food supply, to help enterprises adopt eco-standards, and to improve the investment climate; (iii) expansion of working capital and longer term financing for agribusinesses, with greater emphasis on International Development Association (IDA) and IDA-blend countries; and (iv) increased equity investment in agribusinesses through private equity funds and direct investments. Trade finance was more amenable to rapid short-term expansion.

**Global Agricultural and Food Security Program (GAFSP):** The GAFSP, launched in April 2010, is a grant-based partnership that supports a variety of activities by governments and national and regional organizations designed to enhance agricultural development and food security. The $20 billion financing mechanism was created to manage the G-20's increased support to agriculture and food security. The program is being implemented as a Financial Intermediary Fund for which the World Bank serves as trustee. The GAFSP has a public sector window and a private sector window. Nine donors have pledged $1,245 million to GFSAP by June 30, 2012, $941 million of which is earmarked for the public sector and $268 million for the private sector. The GAFSP private sector window is administered by IFC.

**Agricultural Price Risk Management:** This IFC-led program was announced in June 2011 and aims to provide up to an initial $4 billion in protection from volatile food prices to developing country farmers, food producers, and consumers. This product, the first of its kind, is expected to improve access to hedging instruments to shield consumers and producers of agriculture commodities from price volatility. It will also protect buyers from price rises in food-related commodities, such as wheat, sugar, cocoa, milk, live cattle, corn, soybeans, and rice.

*continued on page 6*

**Horn of Africa Program:** This $500 million program, introduced in July 2011, is designed to assist drought victims. It was scaled up to $1.9 billion in September 2011. The program has short-term and long-term components. While not a direct response to the food price crisis, it targets the same countries that were affected by that crisis and has similar food security, social protection, and agricultural development objectives.

The World Bank Group response program was unique among global financial institutions in articulating a comprehensive, concrete, and fast-disbursing financial support program that provided hard-hit countries with a menu of options for crisis mitigation. Together with the Bank Group's longer-term regular agricultural and social protection programs and knowledge-based policy advice, the GFRP helped solidify the Bank's position as a player on food security matters. The Bank's constructive participation in the UN High-Level Task Force and contribution to G-8 and G-20 meetings helped the international community to outline a coherent international approach to the crisis response.

Design and implementation of the short-term support program helped to build experience for subsequent broader institutional crisis response mechanisms. In the past few years, the Bank Group introduced several new instruments that build on some of the lessons learned through GFRP. These include the IDA Crisis Response Window (CRW) and IDA Immediate Response Mechanism (IRM). Crisis preparedness is now arguably greater, and the Bank Group has more capacity to respond when client demand increases due to unexpected events.

## The Evaluation

This evaluation assesses the effectiveness of the World Bank Group's response in terms of mitigating the short-run impacts of the food price crisis, focusing on developments from mid-2008 to mid-2012. It also assesses progress made by the Bank Group in helping vulnerable countries build resilience to future food price shocks. Based on the analysis, the report draws lessons aimed at enhancing the effectiveness of future crisis related support by the Bank Group and others, and offers recommendations to strengthen the response to future crises.

The evaluation seeks to answer three main questions:

- How did the Bank Group respond to the global food crisis?

- How effectively did the Bank Group help countries address the short-term effects of the food crisis?

- To what extent did Bank Group engagement during and after the crisis help countries enhance their resilience to future food price shocks?

The analytical framework of the evaluation is based on the results chain in Figure 1.2. The evaluation assesses the effectiveness of the Bank Group's crisis response and resilience enhancement in terms of improving outcomes at the country level for the group of vulnerable countries. It focuses on assessing the inputs, outputs, and intermediate outcomes of the results chain. The results chain and analytical framework were also used to evaluate activities in the agriculture and social safety net sectors. Ultimately, both the short-term crisis response and the longer-term enhancement of resilience contribute to poverty alleviation and economic growth, important elements in enhancing food security at the national level (Box 1.2). However, assessment of those long-run outcomes is beyond the scope of the present evaluation, given the relatively short elapsed time since the response was implemented.

## SCOPE

The evaluation conducts two complementary analyses. The first evaluates the immediate response to the 2007–08 food crisis, with particular attention to the Bank's GFRP and the IFC's Global Food Initiative (GFI). The second assesses the Bank Group's medium-term efforts to help build vulnerable countries' resilience to food price shocks in the future by looking at the nature and focus of the agriculture and social safety net portfolios approved since the food crisis, including IFC's risk management initiative. Hence, the evaluation covers the programs specific to the global food price crisis as well as the portfolio influenced by the crisis. The time frame for the assessment of the lending program in the agriculture and social safety net sectors ranges from three years prior to the crisis through June 2012.

The evaluation also reviews how well the Bank Group partnered with donors and other institutions in responding to the crisis, including the United Nations High-Level Task Force on the Global Food Crisis, the Rome-based Committee on Food Security, the Food and Agriculture Organization (FAO), the International Fund for Agricultural Development (IFAD), the World Food Program (WFP), the International Monetary Fund (IMF), regional development banks, private sector organizations, and OECD.

FIGURE 1.2 Analytical Framework for the Evaluation

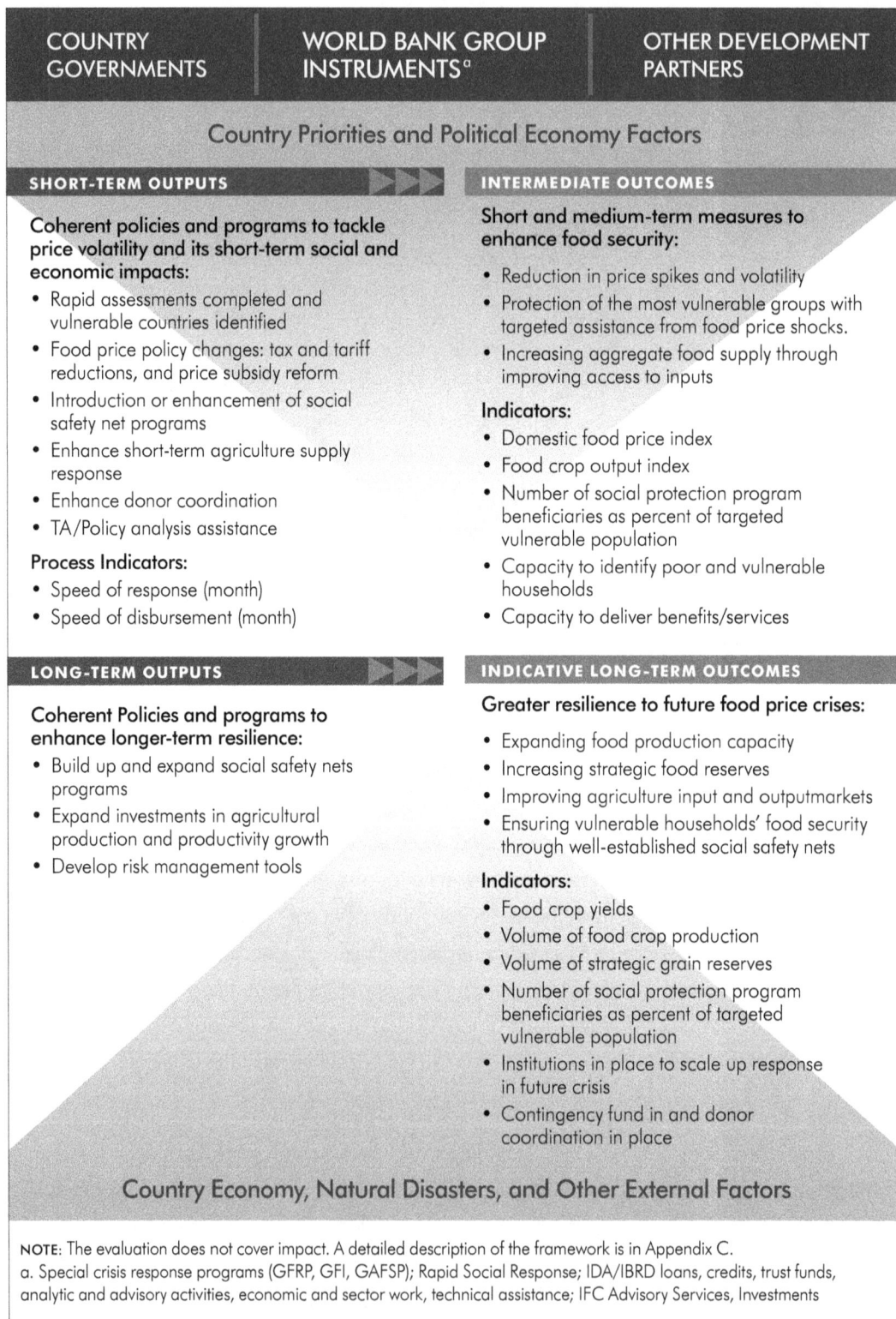

| COUNTRY GOVERNMENTS | WORLD BANK GROUP INSTRUMENTS[a] | OTHER DEVELOPMENT PARTNERS |
|---|---|---|

**Country Priorities and Political Economy Factors**

**SHORT-TERM OUTPUTS** ▶▶▶

Coherent policies and programs to tackle price volatility and its short-term social and economic impacts:

- Rapid assessments completed and vulnerable countries identified
- Food price policy changes: tax and tariff reductions, and price subsidy reform
- Introduction or enhancement of social safety net programs
- Enhance short-term agriculture supply response
- Enhance donor coordination
- TA/Policy analysis assistance

Process Indicators:

- Speed of response (month)
- Speed of disbursement (month)

**INTERMEDIATE OUTCOMES**

Short and medium-term measures to enhance food security:

- Reduction in price spikes and volatility
- Protection of the most vulnerable groups with targeted assistance from food price shocks.
- Increasing aggregate food supply through improving access to inputs

Indicators:

- Domestic food price index
- Food crop output index
- Number of social protection program beneficiaries as percent of targeted vulnerable population
- Capacity to identify poor and vulnerable households
- Capacity to deliver benefits/services

**LONG-TERM OUTPUTS** ▶▶▶

Coherent Policies and programs to enhance longer-term resilience:

- Build up and expand social safety nets programs
- Expand investments in agricultural production and productivity growth
- Develop risk management tools

**INDICATIVE LONG-TERM OUTCOMES**

Greater resilience to future food price crises:

- Expanding food production capacity
- Increasing strategic food reserves
- Improving agriculture input and outputmarkets
- Ensuring vulnerable households' food security through well-established social safety nets

Indicators:

- Food crop yields
- Volume of food crop production
- Volume of strategic grain reserves
- Number of social protection program beneficiaries as percent of targeted vulnerable population
- Institutions in place to scale up response in future crisis
- Contingency fund in and donor coordination in place

**Country Economy, Natural Disasters, and Other External Factors**

NOTE: The evaluation does not cover impact. A detailed description of the framework is in Appendix C.
a. Special crisis response programs (GFRP, GFI, GAFSP); Rapid Social Response; IDA/IBRD loans, credits, trust funds, analytic and advisory activities, economic and sector work, technical assistance; IFC Advisory Services, Investments

People are considered food secure when they have all-time "access *to sufficient, safe, nutritious food to maintain a healthy and active life*" (Definition adopted by the 1996 World Food Summit).

Food security analysts look at the combination of the following three main elements:

**Food Availability:** Food must be available in sufficient quantities and on a consistent basis. It considers stock and production in a given area and the capacity to bring in food from elsewhere, through trade or aid.

**Food Access:** People must be able to regularly acquire adequate quantities of food, through purchase, home production, barter, gifts, borrowing, or food aid.

**Food Utilization:** Consumed food must have a positive nutritional impact on people.

SOURCE: World Food Program website http://www.wfp.org/food-security.

## METHODS

The evidence for the evaluation was distilled from literature and document reviews, semi-structured and in-depth interviews, surveys, program and project analyses, background papers, field-based and desk-based country case studies, and IEG field-based project evaluations (Box 1.3). Some of the evidence is comprehensive—drawing on 100 percent of the lending or analytic and advisory services portfolio—while other evidence is culled from in-depth investigation of purposive samples.

Countries were selected for in-depth assessment based on the amount of support from the GFRP and on geographic region. They include the top five borrowers from the GFRP—Bangladesh, Ethiopia, Nepal, the Philippines, and Tanzania—accounting for 60 percent of GFRP commitments. Ten countries were selected from Africa, the region receiving the most GFRP resources, and two countries were selected from each of the remaining five regions (Table 1.1).

**BOX 1.3** Evaluation Building Blocks

**Desk review of the GFRP Lending Portfolio**, comprising 55 operations in 35 countries. This included an in-depth review of program and project documents to code the type of support and, among the 18 operations that had closed and been rated by IEG, the results achieved. The list of GFRP projects and a summary of the portfolio review are in Appendix D.

**Review of IFC's Food Crisis Response and related Advisory Services.** The evaluation investigated the objectives of IFC's investments in the GFI, capturing dimensions of importance to the food crisis, and the scope and effectiveness of food crisis advisory services.

**Desk review of the Agriculture and Social Safety Net lending portfolios in the World Bank and IFC**, building on the databases and analyses prepared for the two recent IEG evaluations of agriculture and agribusiness and social safety nets. The time frame is FY2006–11, covering 518 agriculture and 200 safety net projects. A summary of the portfolio review for agriculture is in Appendix F and for safety nets in Appendix G.

**Review of Agriculture and Social Safety Net Analytical and Advisory Services.** This covered 891 analytical and advisory service activities (comprising economic and sector work, technical assistance) on agriculture and 289 on social safety nets, approved from FY2006–11.

**Field and desk-based Country Case Studies and IEG Project Performance Assessment Reports.** A total of 20 countries were examined in depth—field-based case studies in nine countries and desk-based reviews in 11 countries—to assess the short-run and resilience building activities associated with the GFRP and the broader World Bank group agriculture and social safety net operations and analytic work. IEG field-based evaluations were conducted for GFRP projects in four of these countries.

**Interviews with officials in partner organizations inside and outside the UN system and with country officials.**

**TABLE 1.1 Distribution of Country Case Studies by Region**

| Region | Field-Based | Desk-Based |
|---|---|---|
| Africa | Kenya | Burundi[a] |
| | Liberia | Ethiopia[a] |
| | Madagascar | Mozambique |
| | Sierra Leone[a] | Rwanda |
| | Tanzania | Senegal |
| East Asia and the Pacific | Philippines | Lao PDR |
| Eastern Europe and Central Asia | Tajikistan | Kyrgyz Republic |
| Latin America and Caribbean | Honduras | Nicaragua |
| Middle East and North Africa | | Djibouti[a] |
| | | Republic of Yemen |
| South Asia | Nepal | Bangladesh |

NOTE: a. A field-based GFRP project evaluation (Project Performance Assessment Report) was also conducted.

## Organization of the Report

The remainder of this report is organized as follows: Chapter 2 assesses the Bank Group's most immediate and extensive short-term response, through the GFRP and IFC's Global Food Initiative. Chapter 3 evaluates Bank Group support in the area of food price policy, market stabilization, and domestic food production. Chapter 4 evaluates the Bank's support to social safety nets. Based on evaluation findings, Chapter 5 draws lessons and provides recommendations for future Bank Group support in responding to food price shocks.

## Endnotes

[1] Price spikes in 2010 were affected by big climate events in temperate exporters, and belief was already widespread that markets had become more vulnerable, so those spikes were more predictable.

[2] These estimates rely on simulation models that tend to overestimate the food price change impact on poverty because of the wage/income effects. Headey (2011) uses results of global Gallup Surveys to argue that the responses of people in many countries suggest that poverty increases due to 2007–08 food price crisis was significantly less than 100 million persons. See Headey and Derek 2011: "Was the global food crisis really a crisis? Simulations versus self reporting." International Food Policy Research Institute (IFPRI) discussion paper 1087.

[3] The Committee on Food Security (established 1974) is a United Nations forum for reviewing and following up on policies concerning world food security. In 2011, an additional organization was created with contributions of some bilateral agencies (the Food Security Cluster) to improve food security responses during humanitarian crises.

## References

Dessus Sebastien, Santiago Herrera, and Rafael de Hoyos. 2008. *"The Impact of Food Inflation on Urban Poverty and its Monetary Cost: Some Back of the Envelope Calculations."* Agricultural Economics 39: 417–29

Ivanic, Maros, and Will Martin. 2008. *"Implications of Higher Global Food Prices for Poverty in Low-Income Countries."* World Bank Policy Research Working Paper Series No. 4594, April 1, 2008.

Tiwari, Sailesh, and Hassan Zaman. 2010. *"The Impact of Economic Shocks on Global Undernourishment."* World Bank, Policy Research Working Paper 5215.

World Bank. 2012. *"Development Policy Lending Retrospective Results, Risks and Reforms."* Washington, D.C.

# 2

# The Global Food Crisis Response Program: Design, Implementation, and Results to Date

## CHAPTER HIGHLIGHTS

- The GFRP, a fast-track loan/grant program mainly targeting hard-hit IDA countries, was launched in May 2008. The operations aimed to mitigate the crisis impact and to foster longer-term resilience to future crises through support for price policy and market stabilization, social protection, and domestic food production and marketing.

- The program committed $1.239 billion for 55 operations in 35 countries; more than half of the support was to four countries—Bangladesh, Ethiopia, the Philippines, and Tanzania. The operations were targeted at countries that were vulnerable to food crises.

- The objectives of GFRP operations were broadly relevant to responding to the food price crisis, but the operations were often not designed to ensure relevance to the country context, in terms of the likely effectiveness of the intervention. Some addressed only longer-term issues.

- On average, GFRP operations were prepared and launched more rapidly than standard Bank operations. However, a third of the emergency operations took three months or more to prepare. There was a trade-off between speed of preparation and quality at entry.

- Food crisis components of ongoing operations funded through additional or supplemental finance were difficult to supervise, monitor, and evaluate because of weak results frameworks at the design stage and their increased complexity, often involving several sectors. The evidence for their effectiveness is poorly documented.

- The performance of two-thirds of the 21 closed GFRP operations was rated moderately satisfactory or better. These projects were prepared and became effective more quickly than the rest of the GFRP portfolio and most closed on time.

- The Bank was one of many international agencies with mandates relevant to the food crisis, including members of the High-Level Task Force on the Global Food Security Crisis and multilateral development banks. Evidence from IEG country visits found that the Bank partnered effectively—though not flawlessly—most of the time at the country level.

This chapter assesses the design, implementation, and performance of the Bank's Global Food Crisis Response Program (GFRP) to date and the Bank's partnerships at the country level in responding to the crisis. The first part of the chapter reviews the GFRP portfolio based on evidence from desk review of design and completion reports, interviews with task team leaders, case studies in the field, and IEG project performance assessments. While only a subset of the 55 operations have closed—the results have been assessed for 21 closed operations to date—much evidence is reviewed on the relevance, design, timeliness, and factors affecting implementation of all approved projects. The second part of the chapter, on partnerships, draws on a review of the international partnership documents and evidence collected during field visits to nine countries.

## Objectives and Design of the GFRP
OBJECTIVES

The GFRP was launched in May 2008 as an umbrella for rapid Bank support to address various aspects of the crisis. The overall framework for the GFRP was prepared by a cross-sectoral team that was a model of collaboration across networks. The program design and its implementation procedures were innovative in the context of the 2008 crisis and provided insights for mainstreaming the Bank's emergency response. It had three objectives:

• Reduce the negative impact of high and volatile food prices on the lives of the poor in a timely manner.

• Support governments in the design of sustainable policies that mitigate the adverse impacts of high and more volatile food prices on poverty while minimizing the creation of long-term market distortions.

• Support broad-based growth in productivity and market participation in agriculture to ensure an adequate and sustainable food supply.[1]

The interventions were to be underpinned by the Bank's existing and emerging analytical and advisory activities and country knowledge, capitalizing on extensive prior efforts in policy analysis and collecting of detailed, periodic, household-level data from a large number of countries. While specific operations would be defined within particular sectoral or subsectoral areas, within a country, all Bank- supported GFRP activities were expected to be nested within an integrated program at the national level and would be compatible with overall Bank Group country programs. As is typical in crises, many other donors and development partners

**BOX 2.1** Major Policy Options Supported by the GFRP

1. **FOOD PRICE POLICY AND MARKET STABILIZATION**

   A. **Food Price Policy: Crisis Options, Transition, and Longer Term Approaches**
      Rapid assessment and analytical support; design of national food policies; information, consultation and participatory advisory services.

   B. **Food Market Stabilization**
      Tax and trade policies; price subsidies on food; grain stock management; price risk management; early warning and weather risk management for food crop production; promotion of bilateral or regional trade, entailing the financing of related technical assistance and infrastructure investments.

2. **DOMESTIC FOOD PRODUCTION AND MARKETING RESPONSE**

   A. **Strengthening Agricultural Production Systems**
      Improving smallholder access to seed and fertilizer; livestock management for vulnerable households; rehabilitation of small-scale irrigation; strengthening farmer access to critical information.

   B. **Reducing Post-Harvest and Marketing Losses**

   C. **Strengthening Access to Finance and Risk Management Tools**
      Improvement and expansion of credit availability to agricultural producers, food processors and traders.

3. **SOCIAL PROTECTION**

   A. **Rapid-Response Diagnostics**

   B. **Short-term Support to the Most Vulnerable Populations**
      Transfer programs (cash transfer, food stamp, food rations); school feeding; public works; nutrition and health programs.

   C. **Strengthening Social Protection Programs**

SOURCE: GFRP Framework Document.

were also likely to be involved, so the GFRP anticipated a need to establish partnerships and coordination mechanisms at the global, regional, and national levels. Within a country, a financing framework would define the roles of all partners, recognizing in particular the responsibilities and comparative advantage of the FAO, UN, and WFP.

The GFRP was initially allocated $1.2 billion and was authorized to operate for three years. On April 16, 2009, the Board increased the funding envelope to $2 billion, but shortened the operating period to two years. It was subsequently extended to June 2012, due to a resurgence of high food prices in the second half of 2010. The total amount of Bank-funded GFRP operations amounted $1.24 billion by the end of FY2012 with 55 operations in 35 countries. In addition to the Bank-funded lending, three externally funded trust funds administered by the Bank provided $344.5 million to food crisis—related activities in 32 countries. Contributors included Australia, Canada, the European Union, Korea, Russia, and Spain. Approval of externally funded projects began only in mid to late 2009, and most were approved well into 2010 and beyond. Evaluating the use of these funds is not yet possible because of the short time since their allocation and limited information availability.

## DESIGN OPTIONS

The GFRP framework paper supported measures in three areas—food price policy and market stabilization, domestic food production and marketing response, and social protection. Within these three broad areas, the framework presented 20 policy options, many of which reflected second-best choices necessitated by the emergency circumstances. It provided country offices, regional staff, and client countries a menu of conceptual entry points for emergency assistance under time pressure (Box 2.1).

## PROCESSING

In light of the need for rapid support, all GFRP operations—whether financed by International Bank for Reconstruction and Development (IBRD)/IDA resources or from the Food Price Crisis Response Trust Fund—were processed as "emergency projects," which have specific guidelines for preparation, appraisal, and approval. They recognize the importance of speedy response by the Bank and therefore exceptions to normal operational procedures are identified (Box 2.2).[2]

The accelerated processing of GFRP operations was to include particular attention to monitoring and evaluation for each activity. Specific objectives, targets, benchmarks, and key performance indicators were to be defined during preparation. Reporting obligations were more stringent than for other operations; implementing agencies were required to issue a midterm progress report 18 months after launch, and a final evaluation report upon completion, to provide details on the accumulated outcomes of the project as well as the main lessons learned.

- Closer engagement between the government and the Bank in the identification and preparation of the operation.

- Accelerated appraisal and review processes, with shortened periods for review and clearance.

- Different balance between ex ante and ex post fiduciary and safeguard requirements and controls, with more requirements confirmed after the operation's approval.

- Substantial amounts (up to 40 percent of the credit) may be allowed as retroactive financing if payments are made by the government up to 12 months before approval.

- Substantial use of additional financing so that a successful relevant ongoing operation can be used as the vehicle for quickly disbursing emergency assistance.

- Simple operations with a short time frame and no effectiveness conditions that are unrelated to the recovery from an emergency.

- Temporary increases in the Bank's cost-sharing percentage.

A Bank Steering Committee oversaw a Secretariat that was established with the responsibility for monitoring implementation of the GFRP and tracking the speed and the results of the operations. The Secretariat reviewed the concept notes for all GFRP operations to ensure that their eligibility.[3] After agreement between the government and the Bank on the concepts and objectives, GFRP operations were appraised and submitted for the Board's "absence of objection." After Board approval of the first two operations, approval was delegated to Bank management. Project documents were circulated to the Board for information, with approval becoming effective five working days thereafter. However, at the request of at least three Executive Directors, the project appraisal documents could be scheduled for Board discussion.[4] A similar procedure was enacted for development policy operations (DPOs). Management was to inform the Board every six months on the status of GFRP implementation. During the period from August 2008 to May 2009, the Bank's President was briefed in writing weekly and the Board monthly on all disbursements and measurable results achieved the previous week by each project.

## The GFRP Lending Portfolio

The GFRP provided rapid assistance to a large number of countries hit by the unanticipated 2008 food price crisis. The projects supported were generally small. By the end of FY2012, the total Bank-funded, Board-approved GFRP program reached $1.24 billion with 55 operations in 35 countries.[5] Of that, $835.8 million was from IDA, $202.4 million was from the Food Price Crisis Response trust fund, and $200 million was from IBRD for a project in the Philippines. The IDA resources went to 17 countries based on available IDA allocations; the bulk of those resources went to three countries—Bangladesh, Ethiopia, and Tanzania. The $202 million trust fund resources were distributed to 27 IDA countries. As a result, GFRP projects were small and of modest scope in many IDA countries.

Most of the GFRP funds were committed quickly. Sixteen of the 55 GFRP operations (although small in volume of funding) were approved in the first three months of GFRP's activities. The bulk of the funds (87 percent) were committed by June 2009, when about two-thirds of GFRP operations were processed. Given the emphasis of GFRP on poorer countries, the bulk of the funding was provided on concessional terms (67 percent IDA and 16 percent grants).

A large share of GFRP support (about 60 percent) benefited Sub-Saharan Africa (Figure 2.1), the most affected region. Higher food prices were a special concern in Africa given that food expenditure accounts for over half of overall household spending, and that Africa imports about half of its rice consumption and about 85 percent of its wheat consumption. Over half of the GFRP operations and commitments (58 percent) were supporting crisis-affected countries in the Africa Region, while other Regions had four to five operations each.

Four countries (Bangladesh, Ethiopia, the Philippines, and Tanzania) received more than half of GFRP's resources (Figure 2.2). The remaining funds were distributed among 31 countries with a large proportion of poor households facing serious food insecurity. In most cases, countries received less than $11 million. Eight countries received almost 80 percent of the funds committed. The modest volume of support received by most countries was due to the limited size of the Food Price Crisis Trust Fund and the limitations of countries' IDA allocations; which are determined by criteria unrelated to vulnerability to crisis effects. In many cases, most of the allocation was already committed to longer-term non-crisis activities that could not be easily restructured. This experience eventually led to the establishment of the IDA Crisis Response Window that the Executive Directors endorsed within IDA16 on February 15, 2011. Using this window, crisis-affected countries could access funds beyond their standard IDA

FIGURE 2.1 Regional Distribution of GFRP Operations and Commitments

GFRP Operations
(n=55)

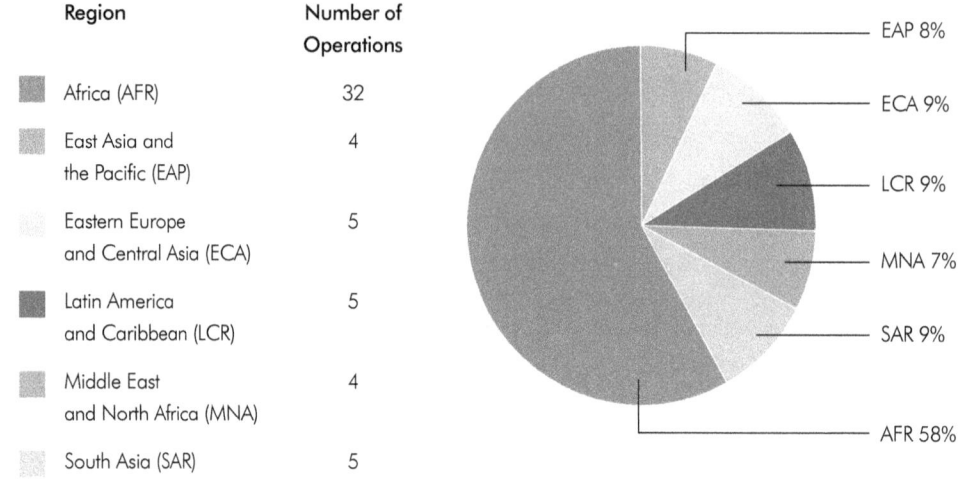

| Region | Number of Operations |
|--------|----------------------|
| Africa (AFR) | 32 |
| East Asia and the Pacific (EAP) | 4 |
| Eastern Europe and Central Asia (ECA) | 5 |
| Latin America and Caribbean (LCR) | 5 |
| Middle East and North Africa (MNA) | 4 |
| South Asia (SAR) | 5 |

EAP 8%
ECA 9%
LCR 9%
MNA 7%
SAR 9%
AFR 58%

GFRP Commitments
($1.239 Billion US)

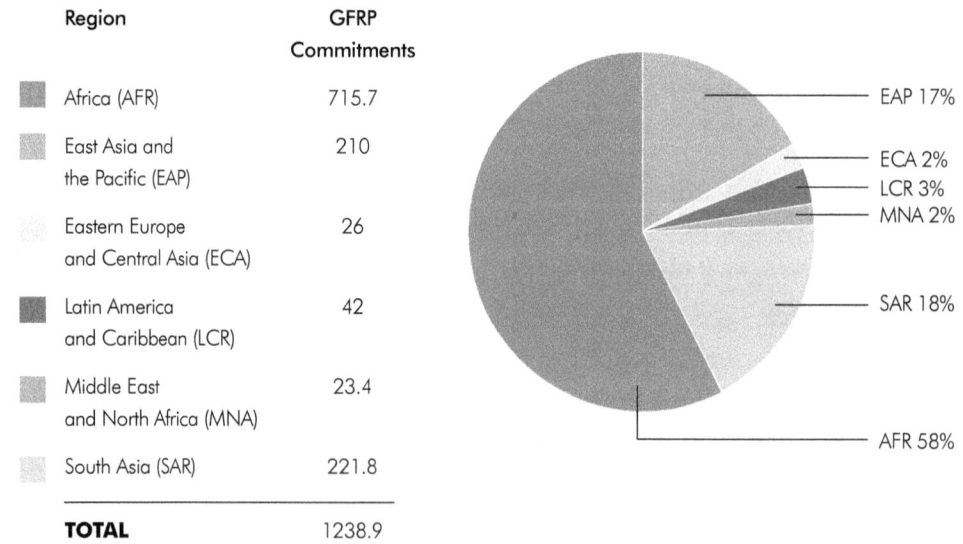

| Region | GFRP Commitments |
|--------|------------------|
| Africa (AFR) | 715.7 |
| East Asia and the Pacific (EAP) | 210 |
| Eastern Europe and Central Asia (ECA) | 26 |
| Latin America and Caribbean (LCR) | 42 |
| Middle East and North Africa (MNA) | 23.4 |
| South Asia (SAR) | 221.8 |
| **TOTAL** | 1238.9 |

EAP 17%
ECA 2%
LCR 3%
MNA 2%
SAR 18%
AFR 58%

SOURCE: World Bank Business Warehouse.

FIGURE 2.2 Distribution of Funds Committed for GFRP Operations among Recipient Economies

Total Loan Amount GFRP Program by Economy

| Economy | World Bank Commitments |
|---|---|
| Ethiopia | 275 |
| Tanzania | 220 |
| Philippines | 200 |
| Bangladesh | 130 |
| Nepal | 83.8 |
| Kenya | 55 |
| Madagascar | 22 |
| Mozambique | 20 |
| Senegal | 20 |
| Nicaragua | 17 |
| Haiti | 15 |
| Burundi | 10 |
| Guinea | 10 |
| Honduras | 10 |
| Kyrgyz Republic | 10 |
| Liberia | 10 |
| Rwanda | 10 |
| Sierra Leone | 10 |
| Republic of Yemen | 10 |
| Benin | 9 |
| Tajikistan | 9 |
| West Bank and Gaza | 8.4 |
| Afghanistan | 8 |
| Central African | 7 |
| Moldova | 7 |
| Niger | 7 |
| Somalia | 7 |
| Togo | 7 |
| South Sudan | 5.7 |
| Cambodia | 5 |
| Djibouti | 5 |
| Guinea-Bissau | 5 |
| Lao PDR | 5 |
| Mali | 5 |
| Comoros | 1 |

ECONOMY

WORLD BANK COMMITMENTS

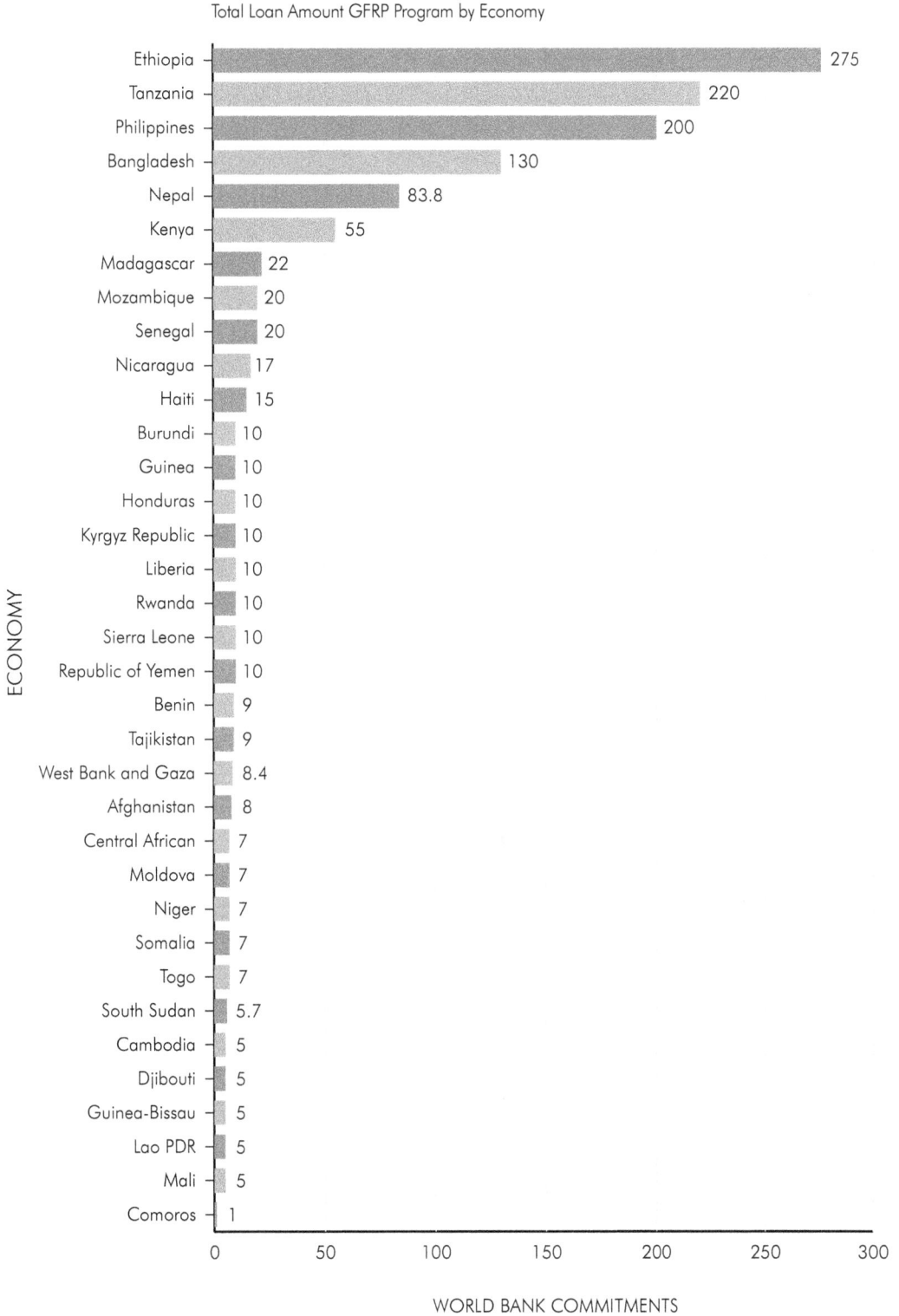

SOURCE: World Bank data.

allocation based on country-specific circumstances (for example, the magnitude of the impact of the crisis, access to alternative financing sources, and ability to finance recovery from the country's own resources).

Ninety-five percent of GFRP funds were allocated to the "most vulnerable" and "vulnerable" countries (Table 2.1). For the purpose of this evaluation, IEG developed a composite index of vulnerability, based on: exposure to global food price increases and whether the country is a food importer or exporter; government capacity to respond; and the size of the vulnerable poor population (see Appendix E for details). Comparing the vulnerability indices for GFRP recipient countries with those for all lower-income and lower-middle-income countries, 28 of the 35 GFRP countries/territories were in the "highly vulnerable" or "vulnerable" groups.[6]

About half of the operations were new and the other half were additional or supplemental finance of new components of ongoing operations. Three-quarters of the operations and two-thirds of commitments were in the form of investment loans, the rest were DPOs (Figure 2.3). By the end of FY2012, nearly half of the GFRP projects had closed, including all of the DPOs.

TABLE 2.1  GFRP Lending by Economy Estimated Vulnerability Level

| Estimated Vulnerability Level | Number of Operations | Number of Economies | Total Commitments | |
|---|---|---|---|---|
| | | | $ Million | Percent |
| Most Vulnerable | 23 | 15 | 370.7 | 29.9 |
| Vulnerable | 21 | 13 | 801.8 | 64.7 |
| Less Vulnerable | 9 | 6 | 58 | 4.7 |
| Other | 2 | 1 | 8.4 | 0.7 |
| TOTAL | 55 | 35 | 1,238.9 | 100 |

SOURCE: World Bank data.
NOTE: Two projects were supported in West Bank and Gaza.

FIGURE 2.3 GFRP Operations and Commitments by Type of Lending Instrument

GFRP Operations

GFRP Commitments in US$ Millions

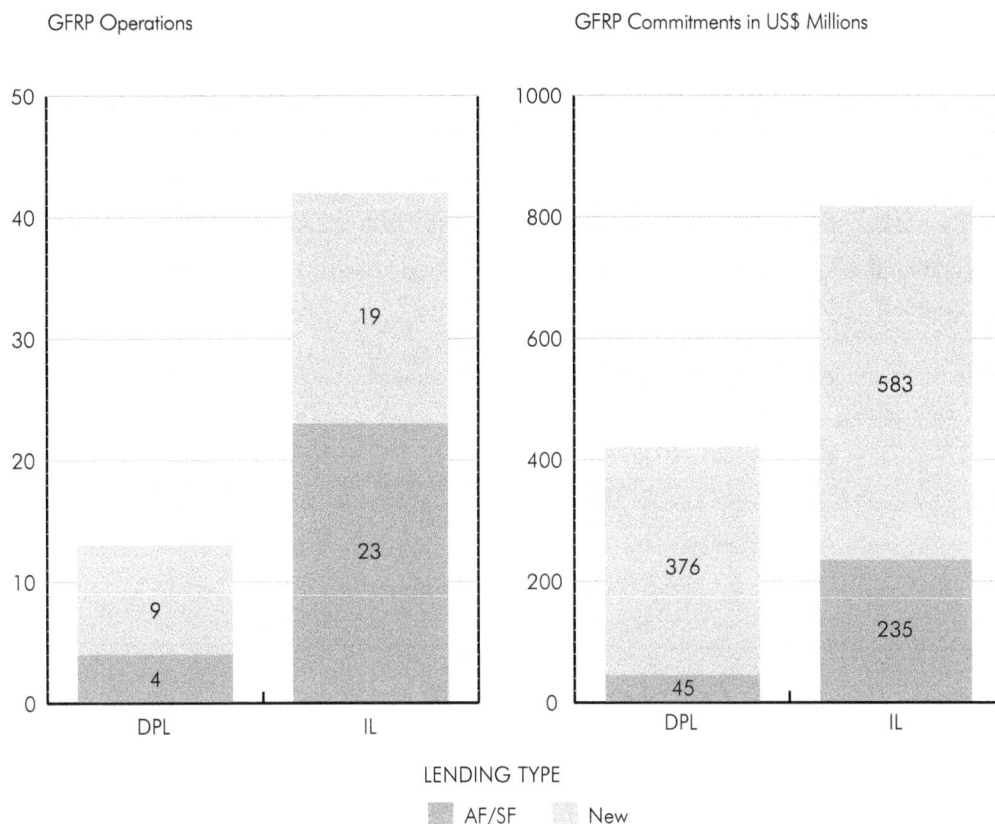

LENDING TYPE

AF/SF     New

SOURCE: World Bank Business Warehouse.

Social protection was the most frequently adopted intervention, followed by support to increase food supply. Most GFRP operations responded to one or more of the program objectives. Of the 55 operations, 18 (33 percent) focused only on social protection, 16 (29 percent) only on increased food supply, 3 (5.5 percent) only on food price policy and macroeconomic stability, and 18 (33 percent) included a mix of objectives. In the latter group, social protection was the most common objective and food production was the most frequently associated objective. While the GFRP Framework presented 20 options, only 9 of these were used (Table 2.2).[7] The most frequently used were improving smallholder access to seed and fertilizer (43 percent of countries), in-kind transfers and food-based programs (38 percent), and public works-employment (18 percent).

Most of the operations addressed the short-term impacts of the crisis; half to nearly two-thirds also addressed longer-term resilience. In 18 of the 20 case study countries in this evaluation, GFRP assistance was relevant to addressing the immediate consequences of the global food crisis. For those operations that focused on social protection, 80 percent addressed immediate

TABLE 2.2 Design Options Adopted in GFRP Operations

| GFRP Activity Areas | | |
|---|---|---|
| Price policy and market stabilization | Social protection to ensure food access and minimize the nutritional impact of the crisis on the poor and vulnerable | Enhancing domestic food production and marketing response |
| Tax and tariff reductions (5 countries) | In-kind transfer and food-based programs (21 countries) | Improving smallholder access to seed and fertilizer (24 countries) |
| Price risk management (2 countries) | Public works—employment (10 Countries) | Rehabilitation of small-scale irrigation systems (5 countries) |
| | Unconditional cash or near cash transfers (6 countries) | |
| | Conditional cash transfers (3 countries) | |
| | General price subsidies (1 country) | |

NOTE: Many GFRP projects financed more than one of the indicative interventions and some countries obtained financing for more than one project. Therefore, some countries are counted more than once. The characteristics of each option and the countries that adopted each are presented in Appendix D.

crisis effects, while 65 percent provided assistance to build up longer-term resilience. For agriculture and food supply-side operations, the shares addressing the immediate and longer-term impacts were 70 and 55 percent, respectively.

Two-thirds of the closed GFRP operations were rated moderately satisfactory or higher on development outcome.[8] Twenty-one GFRP operations had closed and been rated by IEG, of which 15 were freestanding. (The ratings of the six GFRP operations that were additional financing of ongoing projects primarily reflect the performance of the parent project, not the GFRP component.) About two-thirds (67 percent) of the 15 freestanding operations were rated moderately satisfactory or higher on outcome,[9] Bank quality at entry, and Bank supervision. The 21 completed projects that have been assessed were prepared more quickly than the rest of the GFRP portfolio, included almost all of the DPOs, and almost all of them closed on time. It is not clear to what extent the ongoing operations will perform equally well—they took longer to prepare, were more likely to be extended, and were more likely to be investment projects and additional financing for an ongoing project.

# Evaluation Findings

A number of freestanding and supplemental/additional financing operations that created or expanded existing safety nets provided relevant and timely support to mitigate the short-run impact of the food crisis. In Ethiopia, the GFRP contributed additional financing to the ongoing Productive Safety Net Program. The objectives and design of that program were highly relevant to poverty reduction at the time because the increase in food prices had pushed 4.6 million people into severe food insecurity.[10] GFRP additional financing for the Tanzania Second Social Action Fund aimed to improve access of beneficiary households to enhanced socioeconomic services and income generating opportunities. Its objectives and design were relevant to the immediate needs of vulnerable communities. The main objective of the Philippines project, the largest in this group ($200 million), was to help the government address the challenges of the food price crisis. The program's design provided relevant emergency social assistance, establishing a national social protection program in the Department of Social Welfare and Development, guidelines for the food subsidy program, and the adoption and launch of a conditional cash transfers.

The effect of fiscal measures on local food prices depended on the country context. Reductions in domestic taxes and tariffs would be expected to have only a small impact on food prices to consumers when the original tax levels are low, as was the case in many countries in which these policies were supported by GFRP operations. Furthermore, in many low-income countries these taxes constitute a small share of the total retail cost of food, and imported food often is not consumed by the poorest groups. In Burundi the logic behind reducing tariffs and transaction taxes was that, absent those changes, domestic food prices would increase, forcing reductions in food consumption by the poor, increasing hunger, throwing more people into poverty, and increasing social unrest. However, the 13 food staples targeted generally were not imported or consumed by the poor.[11] The Djibouti Food Crisis Response Development Policy Grant, which removed consumption taxes, was not effective for similar reasons. In contrast, under the Philippines GFRP operation, the government agreed to temporarily lift restrictions on private rice importers and at the same time refrained from participating in the international rice market through large contracts. These actions helped to lower domestic food prices and decrease volatility by reducing opportunities for speculation in the global rice market.

Several of the food crisis operations exclusively supported programs for longer-term resilience, many of them in agriculture, with no component to address the immediate effects of the crisis. Projects in Afghanistan, Mozambique, Nepal, Senegal, and Tanzania were designed as part of ongoing longer-term programs of irrigation rehabilitation and not emergency programs.

Therefore, the activities funded by IDA/GFRP had no immediate impact on the food security of poor consumers, as they would not have generated significant increases in food supplies to local markets that would reduce food prices in the short term.

- In Afghanistan, 474 of the 500 small-scale irrigation schemes targeted for the GFRP-funded project had previously been identified for rehabilitation under the National Solidarity Program II, but not funded. Hence, the main impact of the GFRP funding was to increase the scope of an ongoing program, but its impact on food prices in the short term was negligible because the rehabilitation activities were planned to take two years. [12]

- In Mozambique, the $10 million GFRP contribution to the Fifth Poverty Reduction Support Credit was to assist in increasing agricultural production and reducing vulnerability to food price increases. The policy matrix included construction and rehabilitation of irrigation projects, increased access to enhanced production technologies such as seeds, and increased allocations to agriculture in the national budget from 4.7 percent in 2008 to at least 6 percent in 2009 and 7 percent in 2010. [13] As these actions were part of a broader program to support growth and poverty, the GFRP financing was not an emergency response to the food and fuel price crisis but, instead, support for the government budget without any conditions for how the funds would be used or indicators to demonstrate its impact on food security in the short term.

## RESULTS FRAMEWORKS

The quality of the results frameworks for investment projects varied considerably. GFRP project documents were expected to include results frameworks for projects or policy matrices for DPOs. [14] Most documents for GFRP operations provided such frameworks or matrixes. For investment projects, deficiencies, included unclear objectives, absence of indicators, or, when indicators were proposed, uncertain measurability.

The policy matrixes for small DPOs in low-income countries had design weaknesses. Operations with adequate results frameworks had results indicators that were precise and traced a clear causal link between the actions and the anticipated outcomes, where the two were not the same. Using these criteria, only the policy matrixes for the two largest DPOs, the Philippines ($200 million) and Bangladesh ($130 million) operations, were satisfactory. This may reflect the fact that countries that received larger DPOs likely have better monitoring and evaluation systems and that the operations were subject to greater scrutiny (World Bank

2012). In some DPOs, the Bank was severely constrained by donor performance assessment frameworks, and prior actions and results could only be extracted from those frameworks. Among the other weaknesses were the following:

- In the Djibouti Development Policy Grant ($5 million), the expected outcome was vaguely defined as a "somewhat reduced" dependency on food imports and the matrix did not indicate what policies would lead to this reduction, apart from scaling up of fisheries production (a major export from Djibouti).

- The proposal for additional financing for the Honduras Financial Sector Loan had no addendum to the policy matrix to account for the GFRP contribution of $10 million.

- The policy matrix for the Mozambique Poverty Reduction Support Credit 5 (PRSC) offered targets for maize and rice seed production as outcomes for the government's response to the food crisis. These were apparently to arise from an increase in the budget allocation to agriculture. However, the results matrix did not establish a causal chain between the budget allocation, the increased seed production, and the objective of the government's response to the higher food prices: "promoting economic growth as the best means to achieve poverty alleviation."[15] The matrix had neither a target for a food security outcome for the GFRP component nor a specific poverty reduction outcome for PRSC 5.

In many cases, GFRP supplemental funds for DPOs simply augmented funding for the parent operation; few supported policy or institutional reforms. In Honduras and Mali, the supplemental financing merely raised more money for the parent operation. Insufficient attention was given to the impact on food security through either a policy-related reduction in the market price for food, a rapid increase in food supply to the market, or an increase in the real incomes of the poorest of the poor. Finally, the main objectives of most DPOs were to protect core spending on health, education, safety nets, and agriculture and to mitigate the impact of the crisis on the poor, rather than to carry out institutional and policy reforms.

## SPEED OF PREPARATION

Under expedited procedures, GFRP operations were prepared faster than the norm, but in most cases not fast enough for a true emergency. The median processing time for the 55 GFRP operations was 71 days during the period FY2008–11, compared to 236 days for the Bank's entire portfolio of projects during FY2009–11. However, the median masks a substantial range of processing times from 13 days for the Djibouti Food Crisis Development

Policy Grant to 406 days for the Lao PDR Community Nutrition Project. Only 29 percent of the operations were prepared in a month or less; a third took more than three months and three operations took more than a year to prepare (Appendix D).

The GFRP operations in the Africa Region were processed much faster than other projects in the same Region and Bank-wide. Table 2.3 shows the median times elapsed between the concept paper and Board approval and between Board approval and project effectiveness for the GFRP social protection and food supply projects for the Africa Region and for the Bank.

TABLE 2.3 Median Days between Concept, Approval, and Effectiveness for Social Protection and Agriculture Supply Projects (FY2009–11)

| Project Grouping | Time Between Concept Note and Approval (Days) | | Time Between Approval and Effectiveness (Days) | |
|---|---|---|---|---|
| | Social Protection | Agriculture | Social Protection | Agriculture |
| GFRP Projects in the Africa Region | 83 | 81 | 30 | 14 |
| World Bank Portfolio in the Africa Region | 209 | 217 | 127 | 136 |
| All GFRP Projects | 74 | 81 | 32 | 34 |
| All of the World Bank Portfolio | 164 | 238 | 115 | 133 |

SOURCE: World Bank portfolio data, Appendix D.
NOTE: "Mixed" GFRP operations are not included in the Agriculture or the Social Protection GFRP portfolio. The Agriculture/ Social Protection World Bank portfolio is the group of non-GFRP regular projects approved FY2009–11.

Previous evaluations of the Bank's portfolio have confirmed that the quality of a project's design at entry is a major factor in explaining the quality of its outcome. A fast-track approach, while cutting some corners, need not necessarily undermine the quality of the Bank's work because experience and knowledge should make it clear which corners can be cut without undermining quality. To the extent that the GFRP-funded activities provided "additional" or "supplemental" financing and were congruent with the parent project, much of the basic project preparation for these projects and programs had already been done. This was not the case when, for example, a GFRP component was added to an existing project that had little or no affinity with food security or social protection.

GFRP additional financing of activities that were not closely related to the parent project often resulted in mixing of disciplines, which ultimately complicated and undermined implementation. While the overall framework for GFRP operations was designed by a cross-sectoral team, the design and implementation of specific operations was led by single sector units dealing with poverty, agriculture, or social protection. There was cooperation between sector units in some of the case study countries, but not in others.[16] In many operations, the GFRP-financed activities were not closely related to the parent project. For example, there is no record of the specific activities financed or the results of the $10 million in additional financing to the Honduras Financial Sector Credit. Similarly, the additional financing for the Liberia Community Empowerment Project II,[17] a cash-for-work program, augmented funding for a typical community-driven development project focused on small infrastructure and services in rural communities. Finally, the Mozambique the Fifth Poverty Reduction Support Credit was used to carry a GFRP program to improve domestic seed production and growth in the agricultural sector. Grafting the GFRP component to PRSC 5 was relatively simple, but there was no accountability for use of the additional funds or the intended outcome once the funds had become part of the government budget.

GFRP operations were reviewed according to Bank guidelines, but the rigor of the review varied due to quality assurance issues (see endnote 3). A report to the Board on October 7, 2008, stated that the GFRP Secretariat and the Bank Steering Committee (chaired by a managing director and reporting to the Board) reviewed proposals from country teams to ensure a "fit" with GFRP objectives. The report also stated that Bank regional management was using the standard Regional Operations Committee (ROC) process or the Regional Rapid Response Committees (RRRCs) to review projects. A search of project files and documents[18] to assess the quality of project preparation from formulating the project concept and the review processes leading to project approval found energetic due diligence by staff in following guidelines such as "Rapid Response to Crises and Emergencies" (Operational Policy 8.00). However, there were major shortcomings in the substance of the process.

Some GFRP operations were not adequately reviewed during preparation, reducing their effectiveness. For example, the $10 million Burundi operation went from concept paper to approval in 19 days. IEG's Project Performance Assessment Report (PPAR) found that review of earlier analytical work would have led to a conclusion that the proposed tariff and tax reductions would be ineffective. In Djibouti, the $5 million Food Crisis Response Development Policy Grant was processed in 13 days and had similar conceptual weaknesses. The $250 million Ethiopia Fertilizer Project, prepared in 83 days, did not assess the logistical challenges, which put the timely delivery of fertilizer to farmers at serious risk.

Few project records mention peer reviewers or have copies of peer review comments.[19] This suggests that many operations may not have benefited from a deliberate peer review process either because the speed of processing did not allow it or because a large proportion were additional financing, supplementary financing, and restructuring, which are usually the exclusive responsibility of the country director and handled within the country management units. For these activities peer reviews were seldom used even though the GFRP funded components were quite distinct in content from the parent project activities. The absence of peer reviews as part of the quality assurance process should be a major concern for all projects.

The concerns of reviewers were ignored for several projects, with consequences for the outcomes. For example, the Ethiopia Fertilizer Project ($250 million) received many critical comments from senior Bank staff. The minutes of the Operations Committee were not nearly as critical as the comments received but did direct the task team to make specific changes to the appraisal document. The Djibouti DPO was also subject to serious questioning, and some reviewers in the ROC questioned the logic of the Honduras financial sector DPO and requested changes in the documentation.

Design weaknesses were evident in some cases. In Honduras, supplemental GFRP finance was added to the First Programmatic Financial Sector Development Policy Credit a financial sector DPO with virtually no conceptual relationship between the GFRP component and the operation's objectives. In Ethiopia, insufficient attention was given to the adequacy of infrastructure and logistics. The design of the $250 million IDA-funded Fertilizer Support Project aimed to ensure the availability of chemical fertilizers for the 2009–10 production season to smallholders. However, the project used an inefficient government-managed fertilizer procurement arrangement. As a result, farmers did not receive fertilizer until the planting season was over. Further, the design also reversed a decade-long effort to reform fertilizer marketing from a public sector monopoly to a competitive private sector model. In the Central African Republic, closer attention to procurement issues in the preparation and initial implementation period of the agricultural component of the GFRP operation might have avoided significant delays in implementation of the project.

There was some tradeoff between the speed of preparation and the quality of project design and implementation. While the preparation processes followed the guidelines, the speed with which processing was done left little time for serious questions and no time for thoughtful answers. The December GFRP 2010 progress report noted that extra resources for supervision were not provided to balance the quick preparation: "Given the short preparation duration of many of the approved projects, some corners will inevitably have been cut and it is the job of supervision to repair them through intensive supervision. But there is no evidence that any extra budget was allocated to allow for that."[20] A tradeoff between preparation speed and quality was found in 7 of the 20 case study countries.

- In Sierra Leone, the Bank introduced a cash-for-work program by adding a component to the National Social Action Project. The new component was prepared and approved in less than three months and was scaled up quickly, based on minimal experience. The design and implementation both had flaws, such as deficient selection of target groups, limited involvement of communities, and limited flexibility in choice of activities that could be suitable for women. With more preparation time and a more gradual buildup of the program, such problems might have been spotted and resolved.

- In Kenya, a $5 million project providing agricultural inputs to small-scale poor farmers was prepared and appraised rapidly and implemented in just 14 months, but had only a moderately satisfactory outcome because of technical and financial weaknesses and doubts about the targeting efficiency of beneficiaries.

- In contrast, a $50 million conditional transfer program for orphans and vulnerable children in Kenya took about a year to Board approval and will not be closed until December 2013. This certainly does not address the difficulties encountered by the target group in 2008.

**BOX 2.3** Food Crisis Support through a Financial Sector DPO in Honduras: An Example of Limited Relevance of Design

The first phase of the Honduras First Programmatic Financial Sector Development Policy Credit (FSDPC) was approved in 2005, with the objective of strengthening the financial sector to ensure its positive contribution to long-term growth and poverty reduction. In July 2008, Bank staff proposed to the Board GFRP supplemental financing of $10 million to support the government's "commitment to maintain macroeconomic stability and preserve the FSDPC's development objectives while implementing its food crisis response program."[a]

The Board questioned the use of a financial sector operation to address the food. Bank staff argued that supplemental financing would reduce risks facing the financial sector because higher food prices would reduce the real income of households, which in turn would increase the risks posed by nonperforming consumer loans from commercial banks. The banks had built up a substantial portfolio of consumer loans during 2007 and were considered vulnerable to reductions in the real income of their borrowers. The program document said that the supplemental financing would "allow the government to continue responding to the food crisis." The Board approved it, but because of political turmoil, almost five months passed before the government signed the supplemental credit. The funds were disbursed to the budget without conditions on their use other than those already agreed in the ongoing operation's policy matrix. There were no monitoring and evaluation requirements for the supplemental financing.

IEG's review of the Implementation Completion and Results Report (ICR) found that the objective of mitigating fiscal pressures from financial sector losses due to restructuring or closure of failing commercial banks was only partially achieved. The only mention of the GFRP in the ICR is a reference to sources of financing. Supplemental financing of the food crisis program would have had no impact on nonperforming consumer loan portfolios of commercial banks because the food crisis program was focused either on the poorest consumers (who were highly unlikely to be commercial bank borrowers), or supported strategies to increase domestic food supply in the medium to longer term (with no short term benefits for poor consumers). In the end there was no evidence that commercial banks benefited or that the food security of the poor had improved.

SOURCE: Staff analysis based on Honduras country study and other project documents.
NOTE: a. Bank Report No. 44805-HN, Program Document for the Proposed Supplemental Financing Credit for the Honduras First FSDPC, July 24, 2008, page 8.

Freestanding GFRP operations were required to have monitoring and evaluation, yet operations that were grafted on to an ongoing project that had no specific food security objective often had no monitoring and evaluation (M&E). In the latter cases, the M&E arrangements for the parent project or program tended to prevail, and in only a few cases were separate M&E arrangements made for the GFRP operation.

- For example, in 2008 the GFRP provided additional financing of $10 million to the Republic of Yemen Third Social Fund, which was already a $473 million operation and at its close in December 2009 was $694 million. The project document for the additional GFRP financing states that the project objective was to support labor-intensive public works in irrigation areas and related earth works, as well as to support "a national household survey to identify the poorest Yemenis suffering from high food and fuel prices."[21] IEG's review of the project's completion report (ICR) noted serious technical problems with the M&E for this project, and in retrospect, it was unlikely that the relatively few activities financed by the GFRP would receive special M&E attention.

- The latest supervision reports for projects in Liberia,[22] Nepal,[23] and Tanzania[24] do not mention GFRP activities or their performance indicators. In the latter case, the original project targets appear to have remained unchanged after the additional financing was provided.

- In the Kenya Agricultural Input Supply Program, no specific M&E program was established. The project relied on secondary data, but consequently there was considerable uncertainty about the results and a cost-benefit analysis could not be done to evaluate the program's efficiency.

- Similarly, for the Honduras operation, IEG found no evidence in the project documents or in the files to confirm that the country management unit followed through on its pledge to use the country dialogue, stakeholder engagement around analytic and advisory activities, and supervision of relevant on-going projects to engage with the government on M&E, and the ICR for the operation did not mention any crisis-mitigation achievements.[25]

The quality of M&E among closed GFRP operations has been modest and few collected or planned to collect welfare outcomes, like nutritional status, from beneficiaries. Among the 21 GFRP operations that have been rated by IEG, two-thirds were rated modest or negligible with respect to the quality of monitoring and evaluation (Appendix D). Given the speed with which emergency operations were prepared, even where project monitoring may have been planned for and undertaken, there was insufficient time to design adequate assessment of results.

## RISKS

GFRP projects have been weak in addressing the numerous potential risks to achieving results that were anticipated in the framework paper. The potential risks identified included the availability of critical additional resources, capacity of client delivery structures, oversight arrangements, coordination among development partners, leakage in the targeting of beneficiaries, and inadequate component design. All of these risks were relevant, but the operations have been weak in addressing them. For example, incorrect targeting of program beneficiaries has been a problem, as in the Kenya Agricultural Input Supply Program. Because there was no M&E for this project, the targeting errors (errors of inclusion), though acknowledged, have been impossible to quantify. Bank staff estimated that leakages of benefits to those not eligible were no more than 5 percent, but there is no evidence to back up this estimate.

## FINANCIAL MANAGEMENT AND PROCUREMENT

Project documents that IEG reviewed stated the intent to have the necessary financial management expertise in place during implementation, but IEG could find no direct evidence that this commitment was kept in all GFRP operations.[26] GFRP operations processed under the Bank's "Rapid Response to Crises and Emergencies" policy (OP 8.00) were granted the same flexibility as other rapid response operations with regard to financial management and procurement. In 21 GFRP operations rated by IEG, 5 showed adequate fiduciary performance, in 4 it was inadequate, and in 12 there was no information on fiduciary performance.

## SUPERVISION AND TIMELY COMPLETION

While most freestanding GFRP operations were regularly supervised, supervision of GFRP components added to other projects was inadequate. Implementation status reports (ISRs) were regularly prepared for most freestanding GFRP projects, and there is evidence of due diligence by task team leaders and Bank management in reviewing those reports. However, more than half of the GFRP operations were additional financing of ongoing projects that have their own identity and their own supervision reporting. The ISRs and ICRs of those parent projects rarely included assessments of the progress of the relatively small GFRP components.[27]

A third of GFRP operations had to extend the closing date, some by more than a year. Only 25 GFRP operations (45 percent) had closed when this report was prepared, so it is not yet possible to assess the timely completion of the full program. Because of delays in closing dates it will be years before all projects are closed. Eighteen operations—a third of the total—have had their closing dates extended, for 12 operations by a year or more. The extensions range from four to 36 months and average 17 months.[28] As some projects have become effective only after a long delay, this implies that some GFRP projects were unable to provide short-term crisis mitigation assistance, although their activities, if implemented satisfactorily, may be beneficial to poor people in the medium term.

While some of the extensions were due to unpredictable circumstances (in Nicaragua the government changed institutional responsibilities for parent project delivery, necessitating new capacity building and training affecting the GFRP operations), others could have been anticipated but probably not addressed because of the speed with which GFRP operations were prepared. For example, the Central African Republic project experienced substantial start-up problems owing to procurement issues associated with the recruitment of nongovernmental organizations to serve as Financial Intermediary Entities responsible for the implementation of the agricultural supply response component. The recruitment problems resulted in an extension of the closing date by 18 months. In non-emergency circumstances, competent implementation and procurement staff would have had to be appointed before the project could be declared effective.[29]

## Coordination

International partnerships are useful for establishing a coherent, coordinated response to emergencies, but most actions are taken at the country level.[30] This section evaluates the Bank's performance on both aspects of its partnerships. The first part covers the international and country architecture for the crisis response—the mandates and division of labor among the main international agencies involved in agriculture and social protection in developing countries. The second part discusses partnership achievements in 20 countries where case studies were conducted.

### INTERNATIONAL COORDINATION

The launch of the UN Secretary General's High-Level Task Force on the Global Food Security Crisis in the first half of 2008 brought together a number of international agencies, including the World Bank, with mandates relevant to the food crisis.[31] The Bank was represented in the Task Force by a managing director, but the day-to-day liaison was conducted by the GFRP Secretariat. The Task Force turned out to be a multiyear response that is still highly relevant, reflecting continuing global concern about food price volatility and supply availability, especially to vulnerable populations. A Comprehensive Framework for Action, developed by the Task Force, was designed to encourage concerted responses to the food price crisis with actions that respond to the immediate needs of vulnerable populations and contribute to longer-term resilience. The Comprehensive Framework for Action benefited from significant Bank staff inputs and from the analytical effort undertaken in the course of preparing the GFRP. The Framework was updated in 2010 to better reflect ways in which UN System bodies advise the national authorities and numerous other stakeholders engaged in promoting food and nutrition security.[32, 33]

The roadmap developed by the High-Level Task Force was taken up by the G-8 in 2009 (Aquila) and the G-20 in 2009 (Pittsburgh) and 2011 (Cannes), with the latter building on an unprecedented meeting of G-20 Agriculture Ministries.[34, 35] The roadmap incorporates strategies reflected in the Bank's response via the GFRP and the Global Agriculture and Food Security Program (GAFSP), which assisted in the implementation of the pledges made by G-20 members in Pittsburgh.[36]

The Bank's institutional mandate covered all of the nine areas in the Comprehensive Framework matrix, the broadest coverage of any of the partners (Table 2.4). Most of the other agencies participating in the High-Level Task Force were specialized, covering one or two topics from a particular vantage point. In agriculture, for example, FAO and the International Fund for Agricultural Development (IFAD) focus on smallholder agriculture, albeit with broad policy, operational, and research perspectives with which to assess food crisis issues. But for social protection, most listed agencies are more specialized, with each covering one piece of the agenda—from the United Nations Children's Fund's (UNICEF) focus on children's nutrition to the IMF's focus on fiscal costs of safety net programs. The Bank as an institution had the potential to bridge the specialist agencies' expertise.

The Bank's mandate also clearly covers both short-term and resilience-building activities in agriculture and social protection. In contrast, the mandates of the Rome-based agencies are more narrowly focused, with only the WFP focusing on social protection per se. Similarly, the IMF does not count agriculture issues among its areas of competence, counting only the fiscally relevant social protection as core mandates. The international agencies most approaching the Bank's broader institutional mandate are the regional development banks, which are geographically focused and not members of the High-Level Task Force (Appendix B).

However, in implementation of the GFRP and sectoral operations, the Bank was unable to bridge between the many areas in which it had expertise. First, the Bank was not able to translate its institutional universality into operations that effectively bridged agriculture and social protection. Operations that attempted to do so, often for reasons of internal budget constraints, typically fell victim to pressures that impeded working across network boundaries. Second—and quite important in considering institutional comparative advantage for

TABLE 2.4  Institutional Mandates Cited in the Comprehensive Framework for Action Matrix

| | Immediate Availability Food and Nutrition | | | | Longer-Term Food and Nutrition Security | | | | |
| --- | --- | --- | --- | --- | --- | --- | --- | --- | --- |
| | Food Aid | Smallholder Production | Trade and Tax Policy | Macro Impacts | Social Protection Expanded | Small-Holder Sustained | International Food Markets | Biofuel Consensus | Global Information |
| FAO | | X | | | | X | | X | X |
| IFAD | | X | | | | X | | | |
| IMF | | | X | X | X | | | X | X |
| OCHA | X | | | | | | | | |
| OECD | | | | | | | | X | |
| UNCTAD | | | X | | | X | X | X | |
| UNDP | | | X | | X | X | | X | |
| UNEP | | | | | X | X | | X | |
| UNHCR | X | | | | X | | | | X |
| UNICEF | X | | | | X | | | | X |
| WB | X | X | X | X | X | X | X | X | X |
| WFP | X | X | | | X | | | | X |
| WHO | X | | | | | | | | X |
| WTO | | | X | | | | X | | |
| TOTAL WITH MANDATE | 6 | 4 | 5 | 2 | 7 | 6 | 3 | 7 | 7 |

SOURCE: United Nations High Level Task Force on the Global Food Security Crisis.

emergency response—some operations were slow in delivering results on the ground. Many operations were processed quickly—at least relative to the Bank's normal long processing times—but several of them neither disbursed quickly nor got results quickly, and some are still under implementation. To a significant extent, this was because, unlike development assistance organizations like FAO and WFP, most Bank-supported operations are executed by the recipient, and implementation in low-income countries was hindered by limited local institutional capacity.

## COUNTRY-LEVEL COORDINATION

The principles for assessing the Bank's partnerships at the country level derive from the aid effectiveness agenda and the Country Assistance Strategy (CAS) process. The top priorities of the aid effectiveness agenda are country ownership, effective and inclusive partnerships, and results.[37] The Bank scored high on the recent Paris Declaration monitoring survey, the main tool for tracking progress globally on the aid effectiveness agenda, to which 76 developing countries responded—including nine of the 10 countries discussed later in this section.[38] The World Bank Group met or is close to meeting the majority of targets.

The current thinking about good practices for the Bank with respect to aid effectiveness agenda at the country level, as part of the CAS process, is guided by periodic CAS Retrospectives. The 2009 Retrospective highlighted two themes relevant to the response to the food crisis. First, the flexible implementation of country strategies sometimes requires a significant departure from the current CASs in light of the spike in food and oil prices and the global financial crisis. Second, it stressed three aspects of partnership and coordination: spelling out the underlying reasons for the Bank's prioritization and tradeoffs, including how it relates to the activities of donors and other partners; the usefulness of "donor mapping" as an instrument to guide selectivity and achieve "a more effective distribution of aid funds and instruments over the strategic pillars of the government policy;" and ensuring that the division of labor among donors is led by the preferences of the aid recipient, not by those of the development partners.

Drawing on the 20 country studies done for this evaluation (and particularly the nine countries where field visits were conducted), CAS and loan documents, and partner documents, the Bank's food-crisis response activities were owned by the authorities and reflected in the CAS. In half of the nine countries visited, this reflected chronic food insecurity and widespread deprivation that was aggravated by the crisis. Broadly speaking there was coherence between the programs of the Bank and those of other donors—reflecting the vast needs that those programs could only partially address in most cases.

The nine countries visited are very different from each other, with major implications for the range of challenges that the Bank and other donors face in helping those countries respond to the crisis, whether acting alone or in partnership. Country income ranged from a low of $170 and $200 in Burundi and Liberia, respectively, to $2,060 for the Philippines,[39] a major factor affecting performance. Box 2.4 compares aid effectiveness evaluation findings for the extreme ends of the range, Philippines and Nepal. Four of the five lowest-income countries are also fragile states, which present additional challenges.[40]

---

**BOX 2.4** Differences in Partnership Context: Nepal and Philippines

The difference that country context can make for partnerships becomes apparent when one compares the findings of the evaluation of the implementation of the Paris Declaration for countries at different levels of per capita income and development.

- In the Philippines, the evaluation found that the government had "strong leadership in aid coordination with donors...," and rated the "current status of alignment of aid to [government] priorities and country systems ... as 'good to high,'... [with] most of the initiatives and efforts since 2005 ... focused on increasing alignment." Of some relevance to the differences from other countries studied, the evaluation cautions: "while the Philippines is a recipient of aid, it is endowed with capable professionals who are tapped to provide technical assistance to other developing countries under some Third-Country Assistance Programs. Educational institutions in the country also serve as training partners/venues for development practitioners from other developing countries."

- For Nepal, the evaluation found limited capacity within the government, especially at the sectoral level and "little evidence of increased collaboration [among donors] at the operational level outside of the SWAp [Sector Wide Approach] sectors other than in some of the peace related activities." It also found that there had been "some change in joint policy and some joint technical work e.g. the joint country strategy analytic studies done by Asian Development Bank (ADB), the World Bank and the U.K. Department for International Development (DFID), but subsequently they 'went their own way.'"

SOURCE: Second-Phase Country Level Evaluation (CLE 2) of the Implementation of the Paris Declaration (PD) in the Philippines. Final Report. 2011. Joint Evaluation of The Implementation of the Paris Declaration, Phase II. 2010.

Against this background of profound differences across these countries, the evidence from the IEG country visits suggest that the Bank mostly partnered effectively—though by no means "flawlessly"—at the country level. It built on the ongoing progress among donors on the aid effectiveness agenda, with a view to minimizing the strain on the authorities' implementation capacity. In Nicaragua and the Philippines, strong government oversight of donor activities very much shaped what the Bank and other donors did, ensuring (or not) coherence across partners' programs—although this approach sometimes led to frictions among donors, as for example between the Bank and WFP (Nicaragua) about the government-determined geographic division of labor between them on school feeding programs. In Tajikistan, existing donor groups provided effective platforms on which the Bank's and others' response to the food and other crises could take shape. In Liberia and Nepal, however, there was considerable fragmentation across donors and donor programs—especially on safety net programs. In these situations, partners interviewed by the evaluation team reported that the Bank played a constructive role in supporting and adding muscle to the authorities' efforts to establish more coherence across donor-supported programs.

Among the international financial institutions, there was much cooperation, amid some differences on program diagnostics at the country level.

- In its institutional response to the crisis, the IMF deferred to the Bank on agriculture issues and on social protection consistently argued for the narrow targeting of benefits to affected groups—very much in line with the Bank's social protection approach—rather than broad and untargeted subsidies.[41] Eight of the nine countries visited had disbursing programs with the Fund at some time during the food crisis. Nepal and the Philippines were exceptions; both received policy-based quick disbursing support from the Asian Development Bank (ADB).[42] In making the case in 2008–10 for additional financial support in five of the seven countries studied (Kenya, Madagascar, Nicaragua, Tajikistan, and Tanzania), the IMF emphasized the impact of the fuel price rise or the financial crisis as much as or more than the food crisis.

- The World Bank partnered on the country assistance or partnership strategies in three countries where the ADB was active (Nepal, Philippines, and Tajikistan). ADB's financial support to these three countries focused more on the response to the financial crisis than the food crisis. While this might be considered evidence of financial burden sharing and division of labor, there were clearly substantive differences between the two banks in the Philippines, as highlighted in the ADB Independent Evaluation Department's evaluation of the response to the crisis, which included Philippines and Tajikistan among the country case studies.[43]

- The World Bank and the African Development Bank (AfDB) partnered at the country level in six African countries studied (Burundi, Kenya, Liberia, Madagascar, Sierra Leone, and Tanzania), including joint CASs in Liberia and Sierra Leone. More broadly, the AfDB's response strategy aimed to support and complement the activities of other partners, such as the World Bank, and to harmonize in the type of activities supported.[44]

- The only country studied by IEG with Inter-American Development Bank (IDB) involvement was Nicaragua, where the IDB has an extensive program in the social and agricultural sectors. The only country IEG studied with European Bank for Reconstruction and Development (EBRD) support was Tajikistan, where the EBRD involvement did not cover food-related issues.

Coordination was the norm for food and agricultural activities at the individual project/program level, especially with the Rome-based agencies. This was in line with commitments made by High-Level Task Force members about their work in food-insecure countries.[45] For the most part, coordination covered the provision of agricultural inputs—or in the case of WFP, school feeding programs—as the Bank and others provided only limited support for policy reform in the agriculture sector, given the very complex political economy of reform in the sector and the reluctance of authorities to tackle vested interests during the crisis.

The Bank effectively coordinated with the FAO's analytical work in Liberia, Nepal, Tajikistan, and Philippines during the crisis. It provided cofinancing and parallel financing in Kenya, Madagascar, Nicaragua, and other countries. However, there was miscommunication in Nicaragua, where the Bank team was unaware of relevant design features of a parallel FAO-supported project. There was also a controversial cofinancing case in Kenya that covered the distribution of subsidized inputs and that was subsequently extended and enlarged by the European Union.

The Bank and WFP worked effectively in a number of countries, according to two models. In one model, used in Nepal, the Bank scaled up operations developed by WFP and the FAO. In the other model, WFP acted as an implementing agent for Bank-supported activities. Procurement issues were a recurring headache in these latter activities. In a project in Liberia, there were problems with monitoring and evaluation and lack of timely communications about an adverse project development, and in Burundi the Grant Agreement for the GFRP operation was formulated with insufficient precision, hindering enforcement of the full budget that was pledged in the Letter of Development Policy for the implementation by the WFP of the feeding program. The funding shortfall hampered cooperation between the government and WFP, and was one reason for the lengthy delay in implementing the school feeding program.[46]

IFAD and the Bank (and others) partnered on diagnostic work in several countries (including Nepal). With respect to new investment projects during the crisis period, IFAD engaged in Madagascar, Nepal, Philippines, Tajikistan, and Tanzania whether the Bank was also supporting food crisis responses; however, to date, only in the Nepal case is there mention of the activities or Bank coordination with IFAD or on Bank-supported operations.[47]

In most countries studied the social protection landscape was populated with many donors and donor-supported programs seeking to help the poor and the vulnerable, complicating coordination. There were many such donor programs given that nine of the countries were IDA-eligible, five of which were fragile states. In these countries, a common denominator was the school feeding programs pioneered by WFP and used by a number of UN agencies and bilateral donors—and by the World Bank in Sierra Leone as well as in Burundi, Liberia, Nepal, and Nicaragua. The Bank used a different approach in the other four IDA-eligible countries—food for work in Madagascar, social action funding in Tanzania, and support for the beginnings of social protection programs in Kenya and Tajikistan. In some of these cases, the Bank partnered with others and some not, but in all such cases the Bank was fully clued into the donor community supporting safety net operations. The exceptional case in all this was the Philippines, which as a middle-income country was an outlier among GFRP-eligible countries, with a very different profile with respect to implementation capacity, resource availability, and level of development. In this case, the Bank was able to support the authorities' adoption of the new-style social protection approach highlighted in the recent IEG evaluation on social safety nets, as having been pioneered by middle-income countries in Latin America and elsewhere, often with Bank support.[48]

---

### Endnotes

[1] World Bank, 2008, op cit, page (ii), paragraph 6.

[2] GFRP operations were considered "emergency projects," for which the Bank has unique project preparation, appraisal, and approval guidelines. (See Operational Policy 8.00, "Rapid Response to Crises and Emergencies." Development policy operations in post conflict situations are permitted and the conditions are set out in OP8.6, "Development Policy Lending.") The main differences between the guidelines for emergency operations and those for standard Bank operations were: (a) closer involvement between the government and the Bank in the identification and preparation of the operation; (b) the appraisal and review processes are accelerated with shortened time periods for review and clearance; (c) a different balance between ex ante and ex post fiduciary and safeguard requirements and controls with more of these requirements confirmed after the operation's approval; (d) substantial (up to 40 percent of the credit or loan) may be allowed as retroactive financing of payments made by the government up to 12 months before the operation is approved; (e) there is usually substantial use of additional financing in order to use a successful existing relevant operation as the vehicle for quickly disbursing emergency assistance; (f) above all the operations should be simple with a short time frame and there should be no conditions of effectiveness that are unrelated to the recovery from an emergency; and (g) the Bank may agree to temporarily increase its cost sharing percentage for an operation.

[3] It is understood that, while the GFRP Secretariat under the Steering Committee's oversight, reviewed the eligibility of project proposals, it left the details of project preparation, appraisal, review, quality control and implementation to the country management units. There are trade-offs between the level of detail that defines the eligibility of a proposed operation for inclusion in grant-based special programs (such as the Food Price Crisis Respond Trust Fund of the GFRP), the extent of responsibility of the central authorizing unit over quality assurance, and the speed of authorization. Due to the relatively broad eligibility criteria and absence of specific "triggers" for selection, there were many more applicants for the GFRP than there were funds available, but the Secretariat was able to process applications and quickly deny or approve them, in part because it had no responsibility for operational quality assurance. As discussed later in this Chapter, speedy preparation by regional units sometimes resulted in reduced quality at entry, which could have been avoided if additional oversight were to be applied by a central unit, such as Secretariat. While such added scrutiny could slow processing, it could be simplified to a "checklist" and a verification approach rather than through quality assurance.

[4] This is the same practice as is applied to horizontal Adaptable Program Loans.

[5] In addition, under the Externally-Funded GFRP trust funds, there were approved operations totaling $344.5 million.

[6] Twenty-four of those 35 countries are also on WFP's list of vulnerable countries. A similar number were included in the de Janvry/Sadoulet categories of "most vulnerable" or "highly vulnerable" and 21 countries are in the "most vulnerable" and "vulnerable" classes of the Global Food Security Index (GFSI) of the Economist Intelligence Unit. All countries, except Nicaragua and Honduras, were included in at least one of the higher vulnerability lists, and 22 were on three or more lists (see Appendix E).

[7] See World Bank, 2006, Annex 5. The full set of design options is presented in Appendix D.

[8] The project outcome rating assesses the extent to which the operation achieved, or is expected to achieve, its relevant objectives efficiently. For investment operations, the assessment is based on the relevance of objectives and design, achievement of the objectives, and the efficiency of the projects in achieving its objectives. For development policy operations, only relevance and achievement of the objectives are included.

[9] This is on a par with the outcome ratings for Africa Region projects (65 percent moderately satisfactory or higher, 106 projects completed in FY2009–11), slightly better than results for low-income countries (60 percent, 91 projects), and slightly worse than the entire World Bank portfolio (74 percent, 392 projects), which includes middle-income countries. However, because of the relatively small number of closed GFRP projects, comparisons lack statistical significance.

[10] See Ethiopia Country case study.

[11] Bank Report No. 67471-BI, *Project Performance Assessment Report—Burundi: Food Crisis Response Development Policy Grant*, June 14, 2012, page ix.

[12] The ICR states that at the project's close as scheduled in September 2010 rehabilitation 420 schemes were completed and for the remaining 54 schemes block grants had already been disbursed to the bank accounts of communities responsible. There was no M&E program and hence the ICR could not provide data on the increase in the area of land irrigated nor the overall project benefits. However, an estimate of benefits in the ICR was based on similar schemes financed under National Solidarity Program II which resulted in a 32 percent increase in irrigated land area in 2008 and 55 percent increase in productivity. There is no information on how many GFRP-funded irrigation sub-projects were actually operational and delivering increased grain production when the project closed.

[13] Bank Report No. 44846-MZ, *Proposed Credit and Grant to the Republic of Mozambique for a Fifth Poverty Reduction Support Credit*, October 1, 2008, page 57.

[14] Framework Document for Proposed Loans and Grants-World Bank, 2008, Annex 4.

[15] Bank Report No. 44646-MZ, Mozambique: Fifth Poverty Reduction Support Credit, October 1, 2008, page 38.

[16] This finding is consistent with the findings of IEG's 2011 Matrix Evaluation.

[17] GFRP P112083, parent project P105683.

[18] This conclusion is on the basis of a sample of about 85 percent of GFRP operations. Note that for the sample so far about half the records of ROC meetings have been classified as confidential and their content cannot be accessed.

[19] In a significant number of cases it was not possible to ascertain if peer reviewers were involved with the review of operations or not because files of Regional Operations Committee (ROC) meetings were classified as confidential.

[20] Country case studies and staff interviews indicated that in some cases the rapid launch of a GFRP project was due to pressure to demonstrate progress in using the GFRP program.

[21] Bank Report No. 44043-YE, Emergency Additional Financing Grant for the Third Social Fund for Development, June 18, 2008, pages 4 and 9.

[22] GFRP P112083, parent project P104716.

[23] GFRP P114912, parent project P099296.

[24] GFRP P115952, parent project P085786 and GFRP P115873, parent project P085752.

[25] Source: the ROC review of draft project documents.

[26] The framework paper states that for each project or program, "the Bank requires the government to maintain financial management arrangements that are acceptable to the Bank and that, as part of the overall arrangements for implementing the operation, provide reasonable assurance that the proceeds of the loan, credit, or grant are used for the purposes for which it was granted. Minimum internal controls, including internal audit, should be available prior to flow of funds. The recipient country will need to engage the necessary expertise, systems and capacity, or outsource the functions to local consultants or other agencies in the country to work in the project/program implementation unit on fiduciary issues, should this not be available at the outset. On this basis, appropriate financial management arrangements would be designed for each project, which would be consistent with Bank and regional specific requirements, and described fully in each project appraisal document. When available, and considered acceptable to the Bank, the existing country arrangements for fiduciary functions would be used." Where projects were implemented by UN agencies, the framework paper states that "it is preferable to have the UN agencies apply their rules taking into account the understandings reached until the new framework is in place." Framework Paper, page 47, para 164. The "understandings" is a reference to the *"Fiduciary Principles Accord"* dated 5 December 2008 between the World Bank and United Nations on the management of fiduciary issues by UN agencies on behalf of the Bank.

[27] In some the Kenya Inputs Supply Program, implemented in 14 months, supervision was infrequent. However, the program was closely supervised with authorities from the Bank's Nairobi office and separate supervision reports were not considered necessary given the short implementation period. As previously mentioned, this operation had no dedicated M&E system. This severely hampered evaluation of its outcomes. A formal preparation of an ISR would have revealed the meager basis for completing a full ICR and could have stimulated the preparation of more information on the project's progress.

[28] The reasons for the long extensions vary. One group of countries suffered severe political instability (Guinea, Madagascar, and Nepal). Others have been affected by natural disasters. In Bangladesh, floods and a cyclone in 2007 coincided with the food price emergency. In Haiti, the 2010 earthquake created massive disruptions and intensified food insecurity close to Port au Prince. In the Philippines, the GFRP operation received additional financing to deal with damage caused by a typhoon. Both projects in Tajikistan received additional funds from other sources and have been kept going with extensions.

[29] The minutes of the internal Bank review meeting for this project record an acknowledgement that this project was prepared very rapidly.

[30] And consistency between the two levels is clearly essential for effectiveness, as recognized in initiatives and country-based programs. See the Aquila Communiqué, for example, op. cit.

[31] The other institutions included the FAO, IFAD, IMF, the Office for the Coordination of Humanitarian Affairs (OCHA), OECD, United Nations Conference on Trade and Development (UNCTAD), United Nations Development Programme (UNDP), United Nations Environment Programme (UNEP), United Nations Refugee Agency (UNHCR), UNICEF, WFP, and the World Health Organization (WHO).

[32] IEG 2011. Evaluative Lessons from WBG Experience: Growth and Productivity in Agriculture and Agribusiness.

[33] While covering several issues in greater depth, the update "concluded that the [2008] Comprehensive Framework Action's (CFA) analysis and emphases [were] as relevant [then] as they were in 2008 when it was first produced." According to the CFA foreword, it covered a somewhat wider range of issues than the CFA and contained a more detailed "treatment of all aspects of food and nutrition security than its predecessor. It prioritize[d] environmental sustainability, gender equity, the prerequisites for improved nutrition and the needs of those least able to enjoy their right to food. It acknowledges that, while States have the primary role in ensuring food and nutrition security for all, a multiplicity of other actors have vital contributions to make."

[34] See the Aquila Communiqué: L'Aquila" Joint Statement on Global Food Security; and L'Aquila Food Security Initiative (AFSI). Of relevance to the coherence of the international response at the country level, as discussed in subsequent chapters of the evaluation, the communiqué stated, inter alia: "By joining efforts with partners and relevant stakeholders around the world, we can together design and implement an effective food security strategy, with priority on the world's poorest regions. We agree to support a global effort whose core principles are country ownership and effectiveness. We pledge to advance by the end of 2009—consistent with our other actions aimed at improved global governance for food security—the implementation of the Global Partnership for Agriculture and Food Security. Its mission includes enhancing cooperation in achieving global food security, promoting better coordination at the country level and ensuring that local and regional interests are duly voiced and considered. We intend that the Global Partnership will count on a reformed and effective Committee on World Food Security involving all relevant stakeholders, including Governments, International and Regional Organisations, international financial institutions (IFIs), civil society and farmers organizations, the private sector and scientific community.... We support the implementation of country and regional agricultural strategies and plans through country-led coordination processes, consistent with the Accra Agenda for Action and leveraging on the Comprehensive Framework for Action of the UN High Level Task Force and on existing donor coordination mechanisms...."

[35] See the Ministerial Declaration: Action Plan On Food Price Volatility and Agriculture Meeting of G20 Agriculture Ministers. Paris, 2011. http://www.g20.utoronto.ca/agriculture/index.html. Also, The final G20 Communiqué from Cannes included the following passage on agriculture: "Promoting agricultural production is key to feed the world population. To that end, we decide to act in the framework of the Action Plan on Food Price Volatility and Agriculture agreed by our Ministers of Agriculture in June 2011. In particular, we decide to invest in and support research and development of agriculture productivity. We have launched the "Agricultural Market Information System" (AMIS) to reinforce transparency on agricultural products' markets. To improve food security, we commit to develop appropriate risk-management instruments and humanitarian emergency tools. We decide that food purchased for non-commercial humanitarian purposes by the World Food Program will not be subject to export restrictions or extraordinary taxes. We welcome the creation of a "Rapid Response Forum," to improve the international community's capacity to coordinate policies and develop common responses in time of market crises. Improving energy markets and pursuing the Fight against Climate Change." See http://www.g20-g8.com/g8-g20/g20/english/for-the-press/news-releases/g20-leaders-summit-final-communique.1554.html.

[36] See http://www.gafspfund.org/gafsp/content/global-agriculture-and-food-security-program.

[37] See Paris Declaration on Aid Effectiveness, 2005, and Accra Agenda for Action, 2008, at http://www.oecd.org/dataoecd/11/41/34428351.pdf.

[38] Nicaragua was the single case-study country that did not respond to the survey. See World Bank and Aid Effectiveness: Performance to Date and Agenda Ahead. 2011.

[39] The income numbers are for the year 2010, based on the Banks' Atlas Methodology as set out in http://data.worldbank.org/indicator/NY.GNP.PCAP.CD.

[40] The exception is Madagascar, which is not included in the Bank's fragile state ilst. See http://siteresources.worldbank.org/EXTLICUS/Resources/FCS_List_FY12_External_List.pdf.

[41] See IMF Closely Involved in Drive to Relieve Global Food Crisis, IMF Survey Online, May 13, 2008. See also Food Crisis: IMF Backs Some Policy Responses, Voices Caution on Others, IMF Survey Online, May 23, 2008.

[42] The Madagascar program went off track in 2009, because of political developments; the Tajikistan program came on track in 2009, after having been off track for a "non-conforming" purchase.

[43] See *Special Evaluation Study on Real-time Evaluation of Asian Development Bank's Response to the Global Economic Crisis of 2008–2009*. ADB/IED, 2011.

[44] See *The African Food Crisis Response*, African Development Bank, 2008.

[45] According to www.un-foodsecurity.org/countries, the High-Level Task Force (HLTF) identified 30 countries for intense coordination, selected based on the extent of food insecurity and the potential for better results through enhanced interagency coordination. The 30 countries are: Afghanistan, Bangladesh, Benin, Burkina Faso, Burundi, Cambodia, Central African Republic, Djibouti, Eritrea, Ethiopia, Guinea, Guinea Bissau, Haiti, Honduras, Kenya, Lao People's Democratic Republic, Liberia, Madagascar, Malawi, Mali, Mauritania, Mozambique, Nepal, Niger, Pakistan, Rwanda, Sierra Leone, Tajikistan, Tanzania, Timor-Leste, Togo, the Republic of Yemen, and Zimbabwe. That is, 8 of the 10 countries studied by IEG—all but Nicaragua and Philippines—are on the list.

[46] Independent Evaluation Group (2012). Project Performance Assessment Report: Burundi-Food Crisis Response Development Policy Grant, World Bank Report No.67471-BI. Page 11, paras. 3.4, 3.5.

[47] See www.ifad.org.

[48] See *Social Safety Nets: An Evaluation of World Bank Support, 2000–2010*. IEG 2011.

# 3

# Bank Group Support for Agriculture to Mitigate Food Crisis Impacts and Enhance Resilience

## CHAPTER HIGHLIGHTS

- To deal with rising food prices, governments adopted measures to stimulate domestic production, reduce taxes and tariffs, increase consumer and producer subsidies and restrict exports.

- Food price crisis mitigation policies elaborated by Bank's agricultural team as early as 2005 provided a platform for the Bank to provide timely policy advice.

- The World Bank Group developed a framework for its response providing detailed policy advice to governments, recognizing that political and operational constraints in the short run would necessitate adopting second-best policies in some cases.

- Most countries obtaining Bank support for short-term crisis response opted for subsidized input programs, but impact depended on availability of complementary factors and coverage of farmers was limited in most countries, due to limited funds. Thus, any significant aggregate price effects were unlikely in most countries.

- IFC's short-term support focused on expanding agribusiness-related trade finance, working capital and wholesale finance with an increasing share in IDA and IDA-blend countries. Advisory services expanded.

- World Bank agricultural lending expanded significantly after the crisis and is now focused on support to productive agriculture. The volume of agricultural analytic and advisory activities has declined, and is focused more on nonlending technical assistance than on economic and sector work. The decline in sector work has potentially adverse implications for the quality of the agricultural lending program.

- Resilience may improve due to the increased support to agriculture, but portfolio performance has been declining, in part due to staffing and skill mix inadequacies.

The sharp escalation in international food prices in 2007–08 triggered discontent and riots in many developing countries, due to the distress and hardships experienced by vulnerable groups, particularly the poor and near poor. This situation induced many governments to undertake remedial policies aiming to dampen the transmission of higher international food prices to the domestic food market, and induce faster supply response from domestic agricultural producers. The latter objective was made difficult by the simultaneous increase in the international prices of fuel and fertilizers—important inputs for agricultural production—diminishing to some extent the improved incentives that would have come from higher commodity prices.

## Agricultural Crisis Response Policies by Governments

The most common policy response of governments to the higher food prices was to promote domestic production. These policies often involved explicit or implicit subsidies on agricultural inputs such as seeds, fertilizers, or fuel and energy (Table 3.1). Policies that reduced tariffs and taxes also were commonly used in an attempt to mitigate the increase in domestic prices, and universal subsidies on food or on food imports were often introduced for the same purpose. Countries that had strategic grain reserves released them to apply downward pressure on prices, and many food-exporting countries introduced export restrictions (or outright bans). Most of these policies had adverse fiscal implications, and some of them were second-best choices or worse due to absence of targeting or due to the distortions they introduce. The complex trade-offs entailed in the various policy options induced many governments to seek policy advice from the World Bank Group.

TABLE 3.1 Crisis Response Policies Adopted by Governments in 2007–08

| Policy Measure | Percent Adopting |
| --- | --- |
| Promote Domestic Production | 53 |
| Reduce Taxes or Tariffs | 42 |
| Consumer Subsidies/Price Controls | 40 |
| Increase Supply From Public Reserves | 27 |
| Export Restrictions | 27 |

SOURCE: First row is calculated based on Appendix 2 in Benson and others, *Global Food Crises: Monitoring and Assessing Impact to Inform Policy Responses*, Food Policy Report, IFPRI, September 2008. Remaining rows are calculated based on Annex 5 in World Bank, *Addressing the Food Crisis: The Need for Rapid and Coordinated Action*. Paper prepared for the meeting of the Group of Eight Meeting of Finance Ministers, Osaka, Japan, June 13–14, 2008. Washington DC, June 5, 2008.

# Analytical Response to the Crisis by the World Bank Group

As the food price crisis broke, the World Bank was well prepared to provide policy advice on mitigating the effects of the crisis and building resilience in the longer term. Extensive prior analytical and field work by DEC in establishing survey-based household data on consumption and expenditures enabled fast assessment of likely crisis impacts on the poor by staff in DEC and Poverty Reduction and Economic Management Network (PREM) central and regional units. The Agriculture and Rural Development Department (ARD) issued in 2005 a report entitled *Managing Food Price Risks and Instability in an Environment of Market Liberalization*. This report, while mostly providing advice that was relevant for building resilience to food crises (for example, promoting agricultural productivity growth, developing market-based risk management instruments such as futures markets, establishing safety nets), anticipated that there would be occasions requiring short-term interventions, such as utilization of publicly held strategic reserves and manipulation of variable tariffs. It warned that such short-term interventions should avoid undermining long-run market development. Much work on trade policy was also done in the pre-crisis years in the Bank's research department (DEC). Consequently, the Bank was able to produce several reports relatively quickly in the first half of 2008 that assessed crisis causes and impacts, and provided policy advice related to the crisis.[1] In particular, a detailed discussion of policy options and their pros and cons was provided in the framework document for the GFRP.[2] A similar policy discussion is provided in a document prepared by Bank staff as an input to a meeting of the G-8 finance ministers in June 2008.[3] A summary of the Bank's policy advice on price stabilization and agricultural policies is provided in Appendix F.

The Bank's policy advice recognized that domestic political constraints, capacity limitations, and the urgency of the situation would require the implementation of second- or third-best policies in many countries. Thus, while export bans and price controls are considered undesirable, food subsidies are accepted as second best if targeted safety nets cannot be expanded. Similarly, use of strategic reserves to lower prices for all consumers is accepted when better targeting is not possible. Input subsidies are recommended when credit and input markets are underdeveloped, given the long time required to resolve the obstacles in these markets. The discussion carefully points out the trade-offs (in particular, fiscal implications) and risks entailed in different policy interventions.[4]

Regional units provided policy advice through Region-specific reports that were consistent with the general reports.[5] A review of the Regional reports found that aside from a detailed discussion of the onset of the crisis in the Region's countries, and elaboration of the economic and social impacts encountered in those countries, there was a remarkable compatibility in the policy messages between the Regional reports and the more general reports. There were minor variations between the general reports and regional reports in the relative emphases on different measures.[6]

While the global and Regional reports provided consistent general advice, they lacked country-specificity. Among the 20 countries studied, only half had a rapid appraisal of needs and priorities for agriculture and price policies. In some countries, this analysis was undertaken by other donors (particularly FAO), so a Bank-supported analysis was not required. In other countries, the Bank had done a recent sector analysis as part of its regular country work, which provided useful knowledge. But such analyses often were not sufficient to assess the merits or efficacy of crisis mitigation policies undertaken by governments, such as tax and tariff reductions, or new subsidy policies. The absence of such analyses detracted from the usefulness of the crisis response operations supported by the GFRP and the Bank. On the other hand, some GFRP operations entailed studies that improved sector knowledge, as was the case with the CAR project.

The Bank's analytical input was sought in support of the 2008 G-8 meeting, and its global audience expanded in the post-crisis years. The emphasis of crisis-related analytical work shifted to dealing with the risks of future crises and strategies to build resilience and reduce the volatility of global food markets. A paper prepared by a multidepartmental Bank Group team for the Conference on Post-Crisis Growth and Development (Pusan, South Korea, June 2010), analyzed food security challenges globally and advocated coordinated donor support for greater and better investments in agricultural productivity enhancement, and enhancement of vulnerable communities' access to food and nutrition. It highlighted the role of the multidonor-funded Global Agriculture Food and Security Program (GAFSP, for which the Bank Group serves as a trustee).[7] Subsequently, the leaders of the G-20, at their summit meeting in November 2010, requested a number of UN and other agencies, including the World Bank Group, World Trade Organization, and FAO to work with key stakeholders to develop options on mitigating and managing food price volatility. World Bank Group experts were active in the preparation of the multiagency report that was issued in June 2011.[8]

Recent analytical work updates the Bank's policy advice. Further work on higher food prices and their increased volatility was completed recently by the Agriculture and Rural Development Department elaborating the causes, consequences, and policy recommendations, based on the experiences and research over the past few years.[9] The recommendations distinguish between responses to higher price levels and responses to volatility. Poverty-related data and analytical work has been intensified in DEC and PREM, with a new initiative expanding the Living Standard Measurement Survey to focus on agriculture data and on sub-Saharan Africa.

# Short-Term Response in Agriculture

The GFRP-eligible interventions that pertain to agricultural policies and investments fell under two of the three program objectives: *food price policy and market stabilization and enhancing domestic food production and marketing response*. Activities financed under both objectives aimed directly or indirectly to counteract the increase in domestic food prices. All but eight of the 35 countries or national entities receiving GFRP funding had agricultural activities, and some of those eight countries had crisis-response agricultural activities funded by other donors. As indicated in Chapter 2, there were 32 agricultural operations, mostly of relatively small size (less than $6 million). Most governments opted for short-term agricultural supply response enhancements, although some longer-term activities, such as rural infrastructure, capacity building, and irrigation were included (Table 3.2). The total volume of agricultural financing through the GFRP was $668 million, mostly through investment rather than development policy operations. The bulk of GFRP's agricultural assistance went to the Africa Region (79 percent of GFRP agricultural commitments, and 59 percent of agricultural operations).[10]

TABLE 3.2 Activities Supported in 32 GFRP Agricultural Operations

| | Activity | Frequency[a] | Percent |
|---|---|---|---|
| Short-term Impact | Fertilizer/Seed/Input Distribution | 24[b] | 75 |
| | Price Stabilization | 4 | 12 |
| Medium-term Impact | Rural Infrastructure/Facilities | 10 | 31 |
| | Extension | 9 | 28 |
| | Small-scale Irrigation | 8 | 25 |
| | Capacity Building | 5 | 16 |
| | Farmer Organizations Strengthening | 2 | 6 |

SOURCE: Portfolio review.
NOTE: a. Activities are not mutually exclusive.
b. Of these, the program documents for 12 operations state that inputs will be subsidized.

By the time GFRP operations were designed and funded, most recipient governments had already enacted tariff and tax reductions or price control measures. Fifteen of the 20 countries studied had tariff or tax reductions on food items as part of their crisis response policies by the time their GFRP operations were approved. This explains why relatively few GFRP agricultural operations adopted "price stabilization" interventions. While Bank policy advice recommended reducing tariffs and taxes on food staples consumed mostly by the poor, it was also emphasized that these make sense in countries where the tariffs/taxes were high. In countries where the rates were low to begin with, further reductions were not likely to have much of an impact. This was the case in countries like Sierra Leone and Burundi, where the tax reduction was assessed by IEG analyses to have had negligible impact on domestic food prices. But the loss of revenue was disruptive even in countries with low tariffs and taxes. Funding from the GFRP was provided, aiming to allow the government to continue its price stabilization through tariff relief, even though a more thorough analysis would have suggested the relatively minor impact of the policy.

Countries enacted other policies—export bans, price controls, non-targeted price subsidies—in their attempts to stabilize prices that were not generally endorsed by the Bank, but the Bank refrained from imposing conditionalities on these policies in its GFRP agricultural operations. Most likely, this was due to an expectation that such imposition would cause delays in implementation because governments would not be cooperative on these policy issues, given obvious domestic political difficulties for government in reversing crisis response policies. .Additionally, trade policy and price interventions often fall under countries' obligations to the World Trade Organization (WTO), and the Bank has in recent years been taking a cautious approach on such issues so as not to appear to examine the consistency of countries' policies with their WTO commitments, which is a matter within WTO mandate. Evidently, the urgency of immediate financial assistance was given the highest priority. This created inconsistencies between the Bank's declared position and its de facto acceptance of these policies. Furthermore, in countries like Guinea and Tanzania, the export bans, which reduce prices and depresses producer incentives, were contradictory to the objective of the GFRP-support to improve producer incentives through subsidized inputs. Nonetheless, the Bank had to recognize governments' reluctance to allow some of their subsidy-induced supply increases to cross the border rather than help reduce domestic prices.

Some of the larger GFRP-funded development policy operations, supported policy or institutional reforms. There was limited focus on reform in the smaller operations that predominated the GFRP portfolio, but in the $200 million Philippines operation, rice tendering procedures were changed (aiming to reduce import costs), with a greater role for nongovernment enterprises. In Bangladesh, larger numbers of private fertilizer dealers were authorized, improving access to fertilizer for smallholder farmers. Government-implemented subsidized input distribution schemes can inhibit the development of input markets. The GFRP framework document advocated minimizing such negative outcomes by using smart subsidies that use vouchers and private sector traders. This was advice was followed in some GFRP programs (for example, Tanzania) but not in others (for example, Ethiopia). Program documents do not provide information on how distribution of inputs was handled; it is likely that government agencies were used, without involvement of private dealers. However, private distribution networks in some of these low-income countries are very thin, and the need to achieve fast implementation may have dictated the use of non-market distribution.

Input subsidy programs were an appropriate crisis mitigation instrument in many countries in the context the 2008 food price crisis. However, the coverage of the input programs supported by the GFRP varied widely across countries, and in many countries it was rather low, due to limitations on funds availability. Funding depended, at least initially, on the amount that could be provided from the highly rationed single-donor trust fund, the IDA "headroom", and possibilities for restructuring within ongoing projects. Countries like Ethiopia and Tanzania received large GFRP credits/grants, enabling coverage of a significant proportion of their farming population with their input distribution program.[11] At the other extreme, the Lao PDR Rice Productivity Improvement Project targeted 4,000 farmers for support through farmers' groups and the Haiti project aimed to support 6,000 farmers.[12] A review of the available evidence suggests that many GFRP projects with input distribution components benefited (or aimed to benefit) less than 5 percent of smallholder farmers as direct recipients of input or similar in-kind services. These included Guinea, Kenya, Kyrgyz Republic, Nepal, Niger, and Togo. A higher share of the smallholder farming population (5–10 percent) received inputs in the GFRP projects in Benin, the Central African Republic, and Somalia, and an even higher proportion (10–20 percent) were direct input distribution beneficiaries in projects in Nicaragua, South Sudan, and Tajikistan (Appendix F).

Although the support for input distribution likely benefitted farmers, because of the programs' limited coverage it is unlikely that they had a significant impact on raising the short-run domestic food supply or lowering domestic food prices in many of the recipient countries.[13] Many GFRP projects documented increased yields among recipients, increased areas cultivated with more or improved inputs, increased irrigated areas, and increases in aggregate output of staples in project areas (Appendix F). However, other factors affecting these outcomes are not accounted for, making attribution to the GFRP input distribution impossible. The GFRP secretariat indicated that over 5.9 million farm households have been directly reached by GFRP projects (Ethiopia and Tanzania alone would account for over 5 million), and that 529,873 tons of fertilizer (Ethiopia and Tanzania alone account for close to 500,000 tons) and 3,223 tons of seeds have been distributed.[14] However, these aggregates do not necessarily imply country-wide price effects in all the GFRP countries, and no evaluative studies have rigorously established attribution of changes in domestic prices (or prevention of additional increase in prices) to the input distribution schemes supported by the GFRP.

The impact of the subsidized inputs on increasing the food supply (and thereby reducing domestic food prices for all consumers) may have been weakened by the often declared intent to target the subsidy toward smaller and poorer farmers. Maximizing the supply response would suggest targeting the subsidies or inputs to farmers who are likely to generate the largest response. These would often (although not always) be the better-off farmers, who have sufficient resources to acquire the complementary inputs (for example, hired labor, equipment, quality seeds, irrigation water, pesticides) needed to maximize the impact on productivity. In contrast, smaller and poorer farmers may be less familiar with the input, less likely to use it without a subsidy, and have less ability to finance complimentary inputs to raise productivity. A larger supply response may ultimately achieve a larger poverty reduction effect. The best strategy for targeting cannot be determined without a detailed analysis in the context of each country. But this issue was not discussed in the GFRP Framework document, nor is it assessed explicitly in the documents of most GFRP-funded operations.[15]

Input supply operations managed the trade-off between maximizing supply response and targeting the poor in different ways. In Kenya, the fertilizer voucher-based scheme for maize targeted poorer farmers; poverty alleviation for that specific producer group was apparently the main purpose, as farmers were helped in storing their harvest so they could sell their produce months after the harvest, when prices are higher.[16] On the other hand, in the GFRP-funded additional finance to the Togo Community Development Project, the enhanced cereal production component that provided subsidized improved seeds and fertilizers clearly opted to maximize production, regardless of the poverty status of direct beneficiaries.[17] The Tanzania Accelerated Food Security Program (a GFRP-assisted $160 million project) attempted to reconcile the competing objectives of maximizing both output and direct poverty alleviation of participating farmers by targeting the more fertile districts (where farmers are generally better off than in less fertile districts), but distributing the subsidized inputs to poorer farmers within these high-potential districts (which was somewhat incompatible with the objective of maximizing output).[18] The Benin Emergency Food Security Support Project also sought to strike a balance between the competing objectives by earmarking 60 percent of the subsidized fertilizers to bigger producers (over two hectares in the Benin context) while targeting the rest to smaller producers and weaker segments of the farming community. [19]

Most the activities supported under the agricultural supply enhancing component (other than input distribution) have a medium-term horizon, because they have a longer set-up and/or construction time. Small-scale infrastructure, irrigation, extension enhancement, and capacity building are examples. These were permissible within the GFRP framework, but the implicit assumption must have perceived a balance between activities with a short- and long-term supply impact. In the absence of some measures with a short-term impact, it would be difficult to justify the GFRP's expedited "emergency" processing procedures.

In a few countries, the GFRP operations consisted solely or mostly of activities with a medium-term horizon. For example, in Afghanistan, small-scale irrigation was the only activity supported. In Mozambique, the GFRP financed a small proportion of a new Poverty Reduction Support Credit supporting mainly medium-term activities and policy reforms. In Senegal, a GFRP project provided additional finance to an ongoing agricultural operation appraised in April 2010 that aimed to enhance rice production primarily through infrastructure and capacity building. [20] Similarly, additional financing of the Nicaragua Second Agricultural Technology Project, while aiming for improved seed supply to farmers, consisted essentially of capacity enhancement activities, the impact of which will register in the medium term rather than the short-term. In Cambodia, the sustained acceleration of the supply response by smallholders required a longer time to achieve, implying a regular lending operation and a longer implementation period. The rationale for employing the fast processing procedure of the GFRP is less obvious in these cases.

The impact of fertilizer distribution on increased yields depended in many cases on the extent to which complementary inputs, in particular quality seeds, were available in quantities and at the right time to allow farmers to maximize the potential yield. In some countries, it was not possible to secure sufficient supplies of quality seeds, either because the local seed industry was not yet adequately advanced (Ethiopia, [21] Nicaragua) or timely imports could not be arranged. Inadequate infrastructure—such as the transport system in Tanzania[22]—was another factor that hindered maximizing the impact of input distribution as it prevented the marketing of outputs. These constraints are difficult to alter in the short term, yet the project documents do not consistently consider their potential impact the effectiveness of fertilizer distribution.

TABLE 3.3 IFC Net Commitments in Food-Supply Chain (US$ millions)

| Fiscal year | CAGª (1) | Real Sector Non-CAG (2) | Global Finance Programs (3) | IFC Totalᵇ (4=1+2+3) | IDA and Blendᶜ (5) | IDA Ratio (6=5/(1+2)) |
|---|---|---|---|---|---|---|
| 2006 | 675 | 129 | 44 | 848 | 96 | 0.12 |
| 2007 | 816 | 379 | 138 | 1,333 | 141 | 0.12 |
| 2008 | 762 | 17 | 414 | 1,193 | 163 | 0.21 |
| FY06–08 | 2,253 | 525 | 596 | 3,374 | 400 | 0.14 |
| 2009 | 703 | 75 | 758 | 1,536 | 258 | 0.33 |
| 2010 | 530 | 358 | 1,077 | 1,965 | 133 | 0.15 |
| 2011 | 334 | 521 | 1,157 | 2,012 | 124 | 0.15 |
| FY09–11 | 1,567 | 954 | 2,992 | 5,513 | 515 | 0.2 |

SOURCE: IFC data.
NOTE: a. Includes CAG portions of joint ventures with other investment departments.
b. Excludes non-food agriculture and forestry investments and syndications.
c. Does not include global finance programs.

IFC's short-term crisis-response was mostly through expanded trade finance and modest absolute increases in agricultural investments in IDA countries. IFC increased its trade financing operations supporting agribusiness and agricultural trade by 83 percent between FY2008 and FY2009; by FY2010 its trade finance operations had grown by 160 percent relative to pre-crisis levels. It also increased its agricultural investment in IDA and IDA-blend countries by 58 percent at the time of the crisis (between FY2008 and FY2009), but this was from a small base. The actual amount involved is rather small (the increase was a mere $95 million, Table 3.3). IFC's direct investments in food production in Sub-Saharan Africa increased from $37.5 million in FY2008 to $53.7 million in FY2009 and $89.2 million in FY2010.

# Medium and Longer-Term Response
## BANK POLICY ADVICE

Agricultural analytic and advisory activities (AAA), which are critical inputs into the design of effective lending operations, stagnated in the post-crisis period 2009–11 (Table 3.4).[23] These activities are associated with better quality at entry and better outcomes in lending operations.[24] A breakdown of the data indicates that AAA that focus directly on agricultural issues (conducted under the oversight of the Agriculture and Rural Development (ARD) sector) declined sharply in the post-crisis period: Spending dropped from $45 million to $37 million (a decline of 18 percent), and the number of distinct outputs dropped by 21 percent, compared to the pre-crisis period. Trends starting with FY2011 show an upward turn, but the impact of this change will register in upcoming operations. The inadequacy is even more apparent when comparing the average amount of lending per AAA activity, where an increase of 55 percent relative to the pre-crisis years is noted. While there was an increase in agriculture-related non-ARD AAA (both in volume of spending and number of products), this effort is not expected to have the same impact on portfolio quality in the sector as focused country and sector studies.[25]

The decline of analytic and advisory activities is mostly in economic and sector work; nonlending technical assistance increased. There were significant changes in the type of AAA performed (Table 3.5). Economic and sector work declined significantly (particularly in the Africa and South Asia Regions, the largest recipients of agricultural lending), while nonlending

TABLE 3.4 Agricultural Analytical and Advisory Activities Before and After the Crisis

| Period | Number of Agriculture Activities | | | Cost (US$ Millions) | | | Average Cost Per Activity (US$ Thousands) | Average Agriculture Lending Per Activity (US$ Millions) |
|---|---|---|---|---|---|---|---|---|
| | ARD | Non/ARD | Total | ARD | Non/ARD | Total | | |
| 2003–05 | 196 | 273 | 469 | 33 | 14 | 47 | 100 | 16 |
| 2006–08 | 194 | 262 | 456 | 45 | 19 | 64 | 140 | 19.3 |
| 2009–11 | 153 | 282 | 435 | 37 | 28 | 65 | 151 | 29.9 |

SOURCE: World Bank data.

technical assistance increased. Economic and sector work activities are typically formal analytical tasks, subject to well-established quality control procedures. In contrast, technical assistance tasks (to the extent they are analytical) are often less formal, less strategic, and not necessarily subjected to quality control reviews. They tend to focus on narrower issues and are often completed more quickly. Some technical assistance tasks entail supporting specific nonlending activities of clients, which are useful but do not necessarily serve to underpin the World Bank's lending activities.

TABLE 3.5 Economic and Sector Work and Technical Assistance Operations by Region

| Region | FY06–08 | | | FY09–11 | | |
|---|---|---|---|---|---|---|
| | ESW | TA | Total | ESW | TA | Total |
| Africa | 86 | 56 | 142 | 53 | 68 | 121 |
| East Asia and Pacific | 55 | 35 | 90 | 37 | 41 | 78 |
| Europe and Central Asia | 40 | 27 | 67 | 29 | 19 | 48 |
| Latin America and Caribbean | 29 | 3 | 32 | 23 | 16 | 39 |
| Middle East and N. Africa | 20 | 17 | 37 | 16 | 18 | 34 |
| South Asia | 41 | 17 | 58 | 21 | 19 | 40 |
| Regional Studies | 27 | 3 | 30 | 40 | 35 | 75 |
| TOTAL | 298 | 158 | 456 | 219 | 216 | 435 |

SOURCE: World Bank data.
NOTE: ESW=economic and sector work, TA=technical assistance.

TABLE 3.6 Agricultural Lending in Pre- and Post-Crisis Periods by Region

| Region | Lending ARD Sector Board (US$ Millions) | | Lending Other Sector Boards (US$ Millions) | | Number of Operations | | Total Lending | |
|---|---|---|---|---|---|---|---|---|
| | 2006–08 | 2009–11 | 2006–08 | 2009–11 | 2006–08 | 2009–11 | 2006–08 | 2009–11 |
| AFR | 1,451 | 2,755 | 649 | 935 | 83 | 115 | 2,100 | 3,690 |
| EAP | 1,284 | 2,066 | 142 | 415 | 27 | 40 | 1,426 | 2,482 |
| ECA | 934 | 304 | 42 | 123 | 48 | 20 | 975 | 427 |
| LCR | 963 | 1,077 | 259 | 1,341 | 45 | 42 | 1,221 | 2,417 |
| MNA | 200 | 434 | 58 | 115 | 8 | 10 | 258 | 549 |
| SAR | 2,723 | 3,294 | 90 | 139 | 39 | 41 | 2,813 | 3,433 |
| TOTAL | 7,554 | 9,929 | 1,239 | 3,069 | 250 | 268 | 8,793 | 12,998 |

SOURCE: World Bank data.
NOTE: Totals may be off due to rounding.

PATTERNS OF AGRICULTURAL LENDING BEFORE AND AFTER THE 2008 FOOD CRISIS

During the post-crisis period (2009–11), World Bank agriculture commitments climbed. Agriculture-oriented lending grew by 48 percent, from $8.8 billion to $13 billion (Table 3.6).[27] IFC operations in food-oriented agriculture have grown by 63 percent.[28] Indeed, the World Bank's Agricultural Action Plan FY2010–12[29] envisaged a significantly expanded lending and investment program in agriculture, and highlighted five key themes for agricultural operations: raise agricultural productivity, link farmers to market and strengthen value chains, risk and vulnerability, facilitate agricultural entry and exit and rural nonfarm income, enhance environmental services and sustainability.

Most of these areas of focus contribute directly and indirectly to countries' increased resilience in future food crises. By building up the productive capacity and efficiency of agriculture, the ability to produce more food at economically competitive costs is strengthened. Consequently, when world prices increase sharply for reasons external to a country, domestic production can respond, mitigating to some extent the transfer of external price spikes into domestic food markets. This is particularly relevant to otherwise vulnerable poor countries that import a large share of staple consumption. Furthermore, the global expansion of agricultural productive capacity (including in agriculture surplus producer countries) supported by increased agricultural lending can be expected to exert a downward pressure on agricultural commodity prices. This will help to stabilize global food markets and mitigate the potential impact of random supply shocks. In countries where a significant number of the poor are smallholders who are net buyers of food, support to agriculture that includes the smallholder subsector increases their ability to protect their welfare from the impact of food price increases triggered by external price shocks, and improves their ability to take advantage of price increases.

All regions except for Europe and Central Asia had significant growth in Bank-supported agricultural operations (74 percent–113 percent). More indicative of the increased demands on staff and budget resources is the increase in the number of operations, which was highest in East Asian and the Pacific (48 percent) and AFR (39 percent) (Table 3.4). A significant share of the agricultural lending is directed to low- and lower-middle income countries. More than half the Bank's agricultural lending went to the Africa and South Asia regions.

The share of lending supporting agricultural production directly has increased. The bulk of the lending in the post-crisis period is still focused on irrigation/drainage (20 percent) and on general agriculture (20 percent). The share of commitments in operations directly addressing agricultural development, as distinct from non-agricultural operations serving the rural population, rose from 73 percent to 80 percent, a 62 percent increase in the volume (compared to a mere 14 percent increase in the volume of "other" rural lending, Table 3.7. This aspect is conducive to enhanced resilience, as it entails more direct contribution to agricultural productive capacity.

The Global Agriculture and Food Security Program (GAFSP) is based on aid effectiveness principles and is expected to coordinate donor support for strategic, country-led, agricultural and food security plans. It was launched in April 2010 to assist the G-20's support for agriculture and food security to both the public and private sectors. GAFSP finances medium- to long-term investments needed to: raise agricultural productivity, link farmers to markets, reduce risk and vulnerability, improve non-farm rural livelihoods, and scale up the provision of technical assistance and capacity development. The program is being implemented as a Financial Intermediary Fund for which the World Bank serves as trustee. The Bank hosts a

TABLE 3.7 Subsector Composition of Agricultural Lending in Pre- and Post-Crisis Periods

| Subsector[a] | 2006–08 | | 2009–11 | |
|---|---|---|---|---|
| | Lending (US$ Millions) | Share (Percentage) | Lending (US$ Millions) | Share (Percentage) |
| 1. Irrigation/Drainage | 1,786 | 20 | 2,596 | 20 |
| 2. Crops | 329 | 4 | 1,023 | 8 |
| 3. Extension and Research | 700 | 8 | 646 | 5 |
| 4. Animal Production/ Fisheries | 250 | 3 | 327 | 3 |
| 5. Agro-marketing/Trade | 259 | 3 | 754 | 6 |
| 6. Agro-industry | 167 | 2 | 259 | 2 |
| 7. General Agriculture | 1,336 | 15 | 2,642 | 20 |
| 8. Public Administration for Agriculture | 0 | 0 | 816 | 6 |
| 9. Forestry | 441 | 5 | 983 | 8 |
| 10. Rural Finance | 250 | 3 | 50 | 0 |
| 11. Rural Infrastructure[b] | 556 | 6 | 221 | 2 |
| 12. Land Administration | 383 | 4 | 146 | 1 |
| SUBTOTAL (DIRECT AGRICULTURE SUPPORT) | 6,457 | 73 | 10,462 | 80 |
| Other[c] | 2,336 | 27 | 2,535 | 20 |
| TOTAL | 8,793 | 100 | 12,998 | 100 |

SOURCE: World Bank data.

NOTE: a. The commitments in the first nine subsectors reflect components with these codes in all projects where such codes are recorded, regardless of which Sector has oversight over the operation. The commitments in categories 10–11 are in operations handled by the Agriculture and Rural Development Sector.

b. Much infrastructure lending that affects rural areas is under infrastructure sectors and is not reflected in these figures.

c. There are two groups of commitments under the "other" category: (i) Operations that are under the oversight of the ARD Sector but focusing on activities not directly connected to agricultural development (e.g, domestic water supply and sanitation, health, social services). These amounted to $242 million in the period FY2006–08, and $150 million in the period FY2009–11, and (ii) Non-agricultural components in operations under the oversight of the ARD Sector Board whose agricultural components were included in sub-sector categories 1–9 above. These non-agricultural components entail activities that mostly serve the rural population, but do not directly affect agriculture (for example., public administration at central and local government levels other than ministry of agriculture, health and other social services, sanitation and water supply). These amounted to $2,094 million in the period FY2006–08 and $2,385 million in the period FY09–11.

small coordination unit that supports the GAFSP Steering Committee. GAFSP has a public sector window and a private sector window (administered by the IFC). As of June 30, 2012, $1.25 billion had been pledged ($941 million to the public sector window and $268 million to the private sector window, and $40 million remained unassigned), and $752 million had been received. The program had already allocated $658 million to 18 IDA countries through its public sector window and one small grant through its private sector window. The activities are too recent to have been evaluated.

## PATTERNS OF IFC AGRICULTURAL SUPPORT BEFORE AND AFTER THE CRISIS

IFC's Global Trade Finance activities supporting agricultural transactions expanded rapidly. IFC's most rapid response to the food crisis was handled through its Global Trade Finance Program (GTFP). The share of trade finance in IFC's overall agri-supply chain investments grew steadily in the post-crisis years, as core agribusiness investments declined. In addition to the global financial crisis, which brought IFC's countercyclical role to the fore, the decline in direct agribusiness investments in the post-crisis period was due to a deliberate slowing by IFC management as it focused attention on the environmental and social impacts of such investments and devised ways to incorporate the relevant considerations in assessing and managing operations. A manifestation of the related management decision was the self-imposed 18-month moratorium by IFC on its edible oil investments until assessment procedures were revised. The crisis spurred IFC to sharpen its focus on agriculture, leading to a number of new initiatives addressing all elements of the value chain and entailing greater internal coordination and collaboration. In May 2009, IFC introduced the Global Trade Liquidity Program (GTLP), with up to $2 billion in IFC funding and targeted commitments of $4 billion from public sources, supporting $20 billion of the trade transactions of 10 participating banks, including the Africa Export-Import Bank. Phase 2 of the program was introduced in January 2010 with two new components. The GTLP-Guarantee program aims to address the shift in global markets, where banks with improved liquidity positions, face increased risk aversion and lower lending appetite, particularly in Africa. The GTLP-Food and Agriculture program provides short- to medium-term funding. It is designed to extend trade and working capital loans to eligible food and commercial farmers and small and mid-size businesses through regional banks in developing countries in regions with an active food and agriculture export market. Sub-Saharan Africa is the primary target region for this component. GTLP has disbursed a total of $1.8 billion to eight program banks through the end of FY2011 facilitating $8.8 billion of trade transactions globally. Some $1.6 billion (or 18 percent) of the GTLP-supported trade was in Sub-Saharan Africa. GTLP's support for IDA and IDA-blend countries was $2.7 billion (or 31 percent).

A Global Warehouse Financial Program (GWFP) provides additional liquidity to agricultural operators. The GWFP, a $200 million program, introduced in September 2010, aims to increase working capital financing to farmers and agriculture producers in IDA countries by leveraging their production. The program provides banks with liquidity or risk coverage relying on warehouse receipts. GWFP did not commit any funds until April 2012. In December 2011, IFC's Board approved a $2 billion Critical Commodities Finance Program to reduce the risk of food and energy shortages and help maintain stable prices for emerging market buyers. The program is supported both by IFC's own funds ($1 billion) and funds from governments and other development finance institutions, and provides credit for traders and intermediaries that move food and agricultural products in and out of low-income countries. It requires matching funds from participating financial institutions. The program was launched in March 2012.

IFC's direct agribusiness investment strategy shifted toward food exporting countries and Sub-Saharan Africa. From FY2006–08 to FY2009–11, the share of IFC's agribusiness investments received by five large food-producing countries (Argentina, Brazil, Indonesia, Russia, and Ukraine) rose from just under 29 percent to 36 percent. Within this group, a heavy shift took place from the Latin America and Caribbean Region to the Europe and Central Asia Region (primarily to Ukraine). The share of Sub-Saharan Africa rose significantly, from only 2.5 percent (six projects) to 14.9 percent (19 projects).

In June 2011, IFC partnered with J.P. Morgan Chase and Société Générale on a new agriculture risk management product to overcome the market constraints that keep banks from underwriting more price-hedging products, to help increase the use of swaps and forward contracts for corn, wheat, and other commodities. Under a $200 million project, IFC is covering up to 50 percent of the credit risk assumed by J.P. Morgan Chase in hedging instruments. The program aims to make more capital available for agricultural producers and to alleviate banks' country risk and capital constraints, allowing them to meet heavy demand for agricultural-commodity price hedges in emerging markets. Since the exposure associated with risk management operations is typically smaller than the principal amount of hedges made available to clients, these combined credit exposures enable up to $4 billion in price protection to be arranged. The targeted beneficiaries of the program are agricultural producers, consumers, aggregators, cooperatives, and local banks in emerging countries, which have been unable to use such hedging instruments because of high upfront costs and margin requirements.

The Global Index Insurance Facility (GIIF) was established by IFC in December 2009 jointly with the World Bank to expand access to index weather insurance in developing countries.[30] Most of GIIF's activity has consisted of advisory services, financed by the GIIF Trust Fund (GTF), a $34 million multidonor trust fund that provides eligible recipients and beneficiaries with grants to build local capacity, provide financial assistance to GIIF partner institutions, give regulatory policy advice, and support performance-based premiums. In November 2010, GTF awarded two grants in Kenya and one in Rwanda totaling $4.1 million to help expand access to insurance in East Africa.[31] The grants will bring insurance to about 35,000 farmers and 5,000 livestock herders by 2013.

IFC extended advisory services on access to finance by agribusinesses. IFC advisory teams worked with banking and non-banking clients on feasibility studies for specific commodities and supply chains in order to understand their cash flows, profitability, and systemic risks. In the post-crisis period, such activities were carried out in Cambodia, India, Indonesia, Ukraine, and West Africa. The teams have also been building the capacity of client financial institutions in agri-finance (providing training in diagnostics, improving risk management systems and processes, and new product design); linking financial institutions to sustainable supply chains; and promoting access to finance for stakeholders along sustainable supply chains. Based on such an analysis, IFC made a $5 million equity investment in an Indian non-banking finance company (Jain), which is a market leader in the manufacture and distribution of water efficient micro-irrigation equipment. In addition, IFC advisory services has engaged at the firm and sector level to improve farm productivity, energy, and water use efficiency, food safety, and help develop sustainable food supply chains, mainly working with lead firms to train smallholders. Engagement with commodity roundtables has strengthened environmental and social standards in the palm oil, soy, sugar, and cocoa sectors worldwide.

IFC's advisory services to agribusiness increased substantially between 2006–08 and 2009–11. Following its overall strategy regarding the effort to build medium-term resilience in the sector, IFC focused its advisory services on high-productivity exporting countries. Thus, the Europe and Central Asia Region's share in the total advisory expenditure rose, while East Asia's share declined (Appendix F).

IFC's quick response to the crisis was accomplished through multiple trade finance facilities—the Global Trade Finance Program, the Global Trade Liquidity Program, and the Global Warehouse Finance Program. These programs addressed ever-present liquidity and finance constraints inhibiting agriculture-related enterprises due to the higher risks associated with this sector, as perceived by finance institutions. The new IFC-led initiatives promoting price and weather insurance to tackle the implications of these risks, and therefore have a potential to improve access to finance (by changing lenders' perceptions), as well as improving the investment incentives of present and potential operators. However, there is room for enhancing the collaboration between the units handling these operations and IFC's sector departments.

While the increased volume of agricultural lending holds promise and the selected results from some completed projects are encouraging, some important leading indicators, such as the performance of completed projects exiting in FY2009–11, are cause for concern (Appendix F).

The performance of the Bank's agricultural projects exiting in FY2009–11 declined. There has been a general deterioration in the performance ratings of Bank projects completed in the post-crisis period 2009–11. The share of agriculture projects rated "moderately satisfactory" or better on their development outcomes declined by 13 percentage points (from 82 percent in the pre-crisis period to 69 percent in the post-crisis period), a statistically significant drop. The performance of non-agricultural projects also dropped, though by much less (from 77 percent to 73 percent). The risk to development outcome ratings for agriculture projects exiting over this period rose considerably, from 40 percent to 55 percent rated as entailing high or significant risk to development outcomes between the pre- and post-crisis periods. The riskiness of non-agriculture projects also jumped, from 31 percent to 42 percent. The upward trends in both groups of projects are statistically significant. Evidently, the risk to the development outcomes of agricultural projects increased more sharply, and agricultural projects exiting in FY2009–11 entail a (statistically) significantly higher risk than other projects.

In contrast, IEG's three-year rolling average of development outcome ratings indicates no significant change in the success rate of IFC's agribusiness projects. Satisfactory projects were 71 percent of the total in both FY2006–08 and FY2009–11 periods. However, the quality of agribusiness projects relative to overall IFC averages has improved in the FY2009–11 period.[32, 33] IFC's internal Development Outcome Tracking System (DOTS) also shows a significant improvement: of the 50 FY2006–08 projects rated by DOTS, 39 have been rated satisfactory or mostly satisfactory, 11 projects were rated unsatisfactory or mostly unsatisfactory. However, of the 50 FY2009–11 projects rated, 43 were rated satisfactory (including one highly satisfactory) or mostly satisfactory, and only 7 were considered unsatisfactory or mostly unsatisfactory.[34] While the effectiveness of trade finance operations in the agribusiness sector could not be specifically assessed, IEG's analysis of the main trade finance program (GTFP) concluded that it had a high degree of additionality and received positive client feedback on the quality of processing and turnaround time. In subsequent years, the IFC launched innovative programs to expand insurance against agricultural risks.

There are several plausible explanations for the decline in the performance of Bank-supported projects completed in FY2009–11. Many client countries underwent a sequence of food, fuel, financial, and economic crises that stretched fiscal resources and possibly disrupted the work of implementing agencies. Indeed, there was a statistically significant drop in the borrower performance ratings of about 8–9 percentage points in all World Bank projects (including agriculture) completed in the post-crisis period, with only 70–71 percent of projects rated "moderately satisfactory" or better in the post-crisis period.

Factors within the Bank related to circumstances both before and after the crisis may also have adversely affected the performance of projects in the agriculture portfolio. An examination of the ratings of quality at entry of projects exiting in FY2009–11 indicates a statistically significant decline in performance of 13 percentage points (from 72 percent to 59 percent moderately satisfactory or better). There was a similar trend (but a smaller decline) in other Bank projects, whose quality at entry was higher than that of agricultural projects. Since most projects exiting in the post-crisis period were designed and launched before FY2009, the weakness in quality at entry must be attributed to early deficiencies. Two relevant factors were highlighted in the IEG evaluation of agricultural activities, namely, the decline in technical expertise and the inadequacy in analytic and advisory services to update the knowledge base underpinning agricultural operations. These deficiencies continued to characterize the post-crisis period.

IEG has previously flagged the decline in the number of technical specialists in agriculture as an issue. There was a 20 percent decline in specialist skills between 2000 and 2006.[35] The decline in technical specialist ranks continued past 2006, along with an overall decline in the number of agricultural staff (Table 3.8). The numbers have now stagnated at levels that are almost half the number of specialists compared to FY2006, and overall about 20 percent fewer staff. While the numbers of staff were declining and composition of skills worsening,

TABLE 3.8 World Bank Agricultural Staff, FY06–11

|  | Generalists | Specialists | Total |
|---|---|---|---|
| FY06 | 214 | 95 | 309 |
| FY07 | 206 | 59 | 265 |
| FY08 | 186 | 50 | 236 |
| FY09 | 199 | 49 | 248 |
| FY10 | 198 | 47 | 245 |
| FY11 | 199 | 50 | 249 |

SOURCE: World Bank Human Resources data.

the volume of lending has been rising steadily. This seems to have adversely affected quality at entry of older projects, but could just as well affect projects launched in the period FY2009–11.

Quality of supervision ratings for agricultural projects completed in the pre- and post-crisis periods also declined. In the pre-crisis period, the quality of supervision in agricultural projects was similar to that of other Bank projects (87 percent compared to 88 percent moderately satisfactory or better—Appendix F). The quality of supervision of agricultural projects completed in the period FY2009–11 declined by a statistically significant 18 percentage points, with only 69 percent rated "moderately satisfactory" or better. In contrast, the quality of supervision in the other Bank projects declined only slightly (from 86 percent to 83 percent "moderately satisfactory" or better).

The decline in performance on supervision likely reflects the heavy volume of work assigned to the agricultural sector staff during the crisis years 2008–09, when high priority was placed on the fast processing and implementation of crisis response operations. This burden, in combination with not much changed real operational budget, diverted staff attention (and budget resources) from ongoing agricultural projects.[36] In the post-crisis period, the average size of agricultural projects increased considerably, while average real supervision budgets declined by 6 percent (from $114 thousand over FY2006–08 to $108 thousand over FY2010–12). This was most severe in Africa, where budgets declined 32 percent (from $124 thousand to $84 thousand), and in the Middle East and North Africa, where they declined 38 percent (from $124 thousand to $78 thousand).[37] The decline in average supervision resources had likely adversely affected the performance of operations that exited in the period FY2009–11, as well as that of projects that are still ongoing. Staff levels during the period of expanded lending in the post-crisis years have remained relatively unchanged, resulting in high work burdens and deleterious impacts on supervision quality.

The performance reviewed above indicates that the agricultural projects completed in the post-crisis period have a relatively more modest contribution to the enhancement of resilience to future food crises than would be expected based on pre-crisis performance. In the period FY2009–11 there has been an impressive expansion of Bank-supported investments to promote agricultural growth, which could enhance countries' resilience to future food crises. The agricultural operations that were initiated in the post-crisis era are mostly in implementation and their outcomes are not yet known. The review in this section provides only suggestive indications on the likely effectiveness of these ongoing World Bank efforts in terms of projects' development outcome. Nonetheless, the reviews of AAA activities, staff skill mixes, and projects' performance reported above suggest cause for concern and the need for vigilance and greater attention to supervision, and adjustment of human resources and operational budgets to ensure satisfactory performance.

## Lessons from the Bank Group's Agriculture Response

Analytical work is critical in identifying issues and informing both policy advice and financing. The Bank's agriculture analytic and advisory activities have been generally of sound quality, and the lending activities thereby, thus informed had better outcomes than lending activities that were not. However, in some of the poorer IDA countries, such as Ethiopia, Madagascar, and Nepal, little AAA was done in the agriculture sector over several years. IFC advisory services have lacked a focus on relevant agribusiness subsectors. Few advisory services leveraged outcomes by linking with investments (IEG 2011).

Both formal economic and sector work and nonlending technical assistance have a role in underpinning the policy dialogue and the quality of future lending. The evaluation's examination of the composition and regional patterns of analytical and advisory services tasks indicate that overall AAA declined while lending increased. In particular, AAA managed under the ARD Sector, which is more effective compared to that managed by other sectors in supporting agricultural lending, has declined. Within this overall trend, economic and sector work declined significantly (particularly in Africa and South Asia—the largest recipients of agricultural lending), while nonlending technical assistance increased. Economic and sector work is typically a formal analytical task, subject to well-established quality control procedures, and therefore key to improving the knowledge base that underpins policy dialogue and lending operations.

Subsidized fertilizer programs alone are not the solution to the food price crisis. The availability of fertilizer is important to increase crop production. But crop production does not depend on fertilizer alone. Availability of improved seeds is a crucial factor, and inadequate infrastructure, extension, and marketing arrangements limit the effectiveness of fertilizers subsidies. There was not much evidence that aggregate crop production at the national level increased significantly as a result of the subsidized fertilizer programs financed by GFRP.

Input distribution programs that cover only a small share of the farming community (as was the case in most GFRP input supply operations) will not generate a significant supply increase. If the aggregate supply response is not large, it is unlikely that domestic food prices will decline as a result of the intervention. As many of the input supply interventions supported by the GFRP did not have the scale to achieve their price reducing objective, their rationale should have been questioned up-front. The lesson is that input distribution interventions require adequate scale if they are to have welfare effects (through domestic food price reduction) beyond the direct recipients of input support. If the necessary resources for such a scale are not available, or if the interventions are not justified based on a cost-benefit analysis, then the only possible justification for subsidized input distribution is crisis-impact mitigation for recipient poor farmers, and the cost-effectiveness of such a measure compared to other targeted mechanisms needs to be assessed, as well as the targeting strategy (discussed below).

Targeting of input support operations requires a clear strategic focus and careful and transparent monitoring. Input support that aims to generate a macro price-reducing effect should target the most productive farmers where supply response would be greatest. However, in the cases where input support is perceived essentially as a poverty alleviation measure to support poor farmers, who are not likely to be the source of a major supply response due to other constraints they typically face, appropriate mechanisms for targeting need to be employed that minimize the risk of leakage and elite capture. The lesson is that strategic objectives of input support operations need to be carefully thought through prior to their design, so as to weigh the merits of different targeting options.

Regular lending that is directly focused on support to agriculture is conducive to enhanced resilience, as it entails more direct contribution to agriculture productive capacity. The evaluation found that the composition of agricultural lending in the post-crisis period changed relatively slightly, but the share of lending supporting agricultural production directly has increased, as distinct from operations serving the rural population but not directly affecting agriculture, such as rural health, domestic water supply and sanitation, and local government administration. The earlier IEG assessment called for such a reorientation, and the trend should be continued.

Ensuring continuous Bank development effectiveness in agriculture also requires reversing the ongoing deterioration in the sector's portfolio performance. The performance of Bank agricultural projects exiting in FY2009–11 declined. The present evaluation notes a general deterioration in the performance ratings of Bank projects completed in the post-crisis period 2009–11. The decline is more remarkable in the agricultural projects cohort and it is accompanied by a significantly increased risk to the development outcomes. While a decline in borrower performance explains some of the deterioration, World Bank Group—related factors are also likely at fault, such as inadequate economic and sector work (ESW), lacking supervision resources, insufficient staff numbers and mismatched staff skills. These require rectification; otherwise, the ongoing agricultural portfolio may be afflicted by the same declining performance.

# Endnotes

[1] World Bank. Rising Food Prices: Policy Options and World Bank Response, April 9, 2008. Donald Mitchell (2008) 'A note on rising food prices' (mimeo); Ivanic, M. and W. Martin. Implications of Higher Global Food Prices for Poverty in Low-Income Countries. World Bank Policy Research Working Paper Series No. 4594., April 1, 2008.

[2] World Bank. 2008. *Global Food Crisis Response Program.* Washington D.C.—The Worldbank. http://documents. worldbank.org/curated/en/2008/06/9618171/global-food-crisis-response-program.

[3] World Bank. *Addressing the Food Crisis: The Need for Rapid and Coordinated Action.* Paper prepared for the meeting of the Group of Eight Meeting of Finance Ministers, Osaka, Japan, June 13–14, 2008. Washington D.C., June 5, 2008.

[4] World Bank policy recommendations for crisis mitigation were criticized by leading agricultural policy scholar Peter Timmer in an article in Food Policy in 2010. A detailed examination of the objections suggests that the differences of opinion are not as significant as perceived, and stem from referring to a particular Bank publication that has not elaborated on the various nuances of WBG recommendations. See Annex C to this paper.

[5]  a. *Rising Food Prices in East Asia: Challenges and Policy Options* by Milan Brahmbatt and Luc Christiaensen, Report no.44998, May 1, 2008.
   b. The Food Crisis: Global Perspectives and Impact on MENA—Fiscal and Poverty Impact . Ruslan Yemtsov, MNSED. June 16, 2008.
   c. Improving Food Security in Arab Countries. January 2009. WB. IFAD. FAO
   d. High Food Prices: Challenges and opportunities for ECA countries. Asad Alam, the World Bank, June 24, 2008.
   e. Rising Food and Energy Prices in Europe and Central Asia. Report No. 61097, April 2011.
   f. Rising Food Prices. The World Bank's Latin America and Caribbean Region Position Paper. June 2008.
   g. Rising Food Prices in Sub-Saharan Africa: Poverty Impact and Policy Responses. Quentin Wodon and Hassan Zaman, WPS #4738. October 2008.
   h. Food Price Increases in South Asia: National Responses and Regional Dimensions, South Asia Region. World Bank. June 2010.
   i. High Food Prices: Latin American and the Caribbean Responses to a New Normal, World Bank, Latin American and Caribbean Region. 2011.
   The list above focuses on reports that discuss (*inter alia*) policies related to price stabilization, price subsidies and agricultural development. There were additional reports regional that focused solely on safety net and social protection issues. These are referred to in Chapter 4.

[6] One area in which some differences in advice were noted is on the advisability of using strategic grain reserves to stabilize market prices. While the general reports caution that the experience with this tool has been mixed and advise that its use be undertaken within a framework where there are predictable and well-advertised rules, the report for the Middle East and North Africa Region suggests totally refraining from using stocks as a price management tool.

[7] Christopher Delgado, Robert Townsend, Iride Ceccacci, Yurie Tanimichi Hoberg, Saswati Bora,Will Martin, Don Mitchell, Don Larson, Kym Anderson, and Hassan Zaman. *Food Security: The Need for Multilateral Action.* Prepared by a multi-departmental World Bank team for presentation at the Korea-World Bank High Level Conference on Post-Crisis Growth and Development, May 2010.

[8] "Price Volatility in Food and Agricultural Markets: Policy Responses" http://www.oecd.org/agriculture/ pricevolatilityinfoodandagriculturalmarketspolicyresponses.htm.

[9] World Bank. Responding to higher and more volatile world food prices. Washington D.C., May 2012. Report 68420. http://documents.worldbank.org/curated/en/2012/01/16355176/responding-higher-more-volatile-world-food-prices.

[10] It should be noted, however, that the GFRP-funded agricultural operations in Ethiopia and Tanzania alone claimed about two thirds of GFRP's agricultural commitments.

[11] Ethiopia delivered 427,000 tons of fertilizers to more than 3 million farm units (about 37% of the total); Tanzania distributed fertilizer vouchers to close to 2 million farmers (almost half of its farming households).

[12] The Haiti target was reduced to 3,000 farmers after the earthquake due to restructuring, which amounts to less than one percent of farm households. According the latest supervision report, 300 farmers have received the subsidy by March 2012.

[13] Denote P as price, $\Delta$P change in price, y yield, $\Delta$y change in yield, n is the number of farmers receiving inputs, N the total number of farms in the country, and $\Sigma$ the price elasticity of demand. It can be shown that in a simplified partial equilibrium model, where all farms are identical, that $\Delta P/P = (\Delta y/y) \bullet (n/N)/\Sigma$ The elasticity of demand for staples is around -.5, and the yield effects of a seed/fertilizer package under smallholder conditions are quite variable (7% in Ethiopia, 11% in Cambodia, 17.5% in Tajikistan, 32% in Nicaragua, 33% in Tanzania, 116% in Niger, 38%–100% in Somalia, and 30%–133% in the Kyrgyz Republic—see Annex B). Using the formula above, a calculation can show that if a 15% reduction in grain prices (compared to the 30–40 increase during the crisis experienced) is the target, a country expecting a 100% yield increase needs to reach 7.5% of its farming community with an input distribution program, while a country expecting a 25% yield increase needs to reach 30% of its farmers to achieve the target price reduction.

[14] http://siteresources.worldbank.org/EXTSDNET/Resources/Results2012-SDN-Food-Crisis.pdf.

[15] The trade-off is highlighted by other evaluations of responses to the food price crisis (See Keats, S., S. Wiggins, and E. Clay. International rapid responses to the global food crisis of 2007–08. Overseas Development Institute, London, UK, September 2011, P. 25).

[16] Interviews conducted in Kenya as part of this evaluation confirmed that alleviating the poverty of small farmers in selected districts was the key motivation for the GFRP operation, rather than reducing domestic prices. The government policies are aimed at maintaining high producer prices, benefiting mostly the larger producers, who account for two percent of the farms and over 50% of the marketed maize (See World Bank, *Kenya Economic Outlook,* December 2009, pp. 11–14). Within this context, the market impact of assisting relatively small groups of smallholders was not realistically expected to be significant.

[17] "In order to maximize the impact of this program, beneficiaries will be selected on the basis of their productivity, with priority given to the 10 most productive farmers in each producer association."
World Bank. 2008. *Togo—Community Development Project: additional financing.* Washington D.C., page 8. http://documents.worldbank.org/curated/en/2008/09/9929994/togo-community-development-project-additional-financing.

[18] As it turned out, the Kenya projects had leakages of the subsidy to non-eligible farmers (larger and better-connected, i.e., "elite capture") even though the distribution was mediated through community institutions. See World Bank, Kenya Agricultural Input Supply Project: Implementation Completion an Results Report (ICR000148), December 29, 2010, Annex 5, paras. 5–7. While the ICR estimated that the extent of leakage was small, about 2%–3%, IEG's ICR Review pointed out that there was no evidence that the estimate was based on a statistically valid random sample.

[19] World Bank. 2008. *Benin—Emergency Food Security Support Project.* Washington D.C., page 14. http://documents.worldbank.org/curated/en/2008/10/9961933/benin-emergency-food-security-support-project.

[20] The small component within this project that included input distribution to farmers amounted to a mere 6 percent of project cost. World Bank. 2010. *Senegal—Additional financing for Food Security (GFRP) Project: emergency project paper.* Washington D.C. Annex 4. http://documents.worldbank.org/curated/en/2010/04/12204532/senegal-additional-financing-food-security-gfrp-project-emegency-project-paper.

[21] World Bank. *Progress Report—Global Food Crisis Response Program.* Washington D.C., December 29, 2010. P.51.

[22] World Bank. Tanzania—Accelerated Food Security Program. Washington D.C. 2009 p. 5.

[23] IEG's evaluation of Bank Group activities in agriculture pointed out that such activities declined in the years 2006–08. In its response to the IEG evaluation, Bank Group management acknowledged the decline and projected that the inadequacy was actually worsening in the post-2008 period (IEG 2011, p. xviii).

[24] Independent Evaluation Group, World Bank (2011). Growth and Productivity in Agriculture and Agribusiness. Evaluative Lessons from the World Bank Experience. Washington, D.C.: World Bank (pp. 20–21). Analysis by the Rural Policies Thematic Group found that AAA has a significant positive impact on the quality of lending, quality of projects at entry and quality of outcomes. Further findings suggest that ARD sector-focused AAA has a significantly larger impact on ARD portfolio quality and outcomes than regional or global AAA or non-sector specific (multi- or cross-sectoral) AAA (World Bank, Agricultural Action Plan 2013–15, paragraph 152).

[25] Ibid.

[26] Analysis of spending on AAA activities reveals similar patterns and is not presented.

[27] These figures do not include IFC activities. The cohort of lending operations related to agriculture used in the analysis of the Global Food Crisis Evaluation follows the definition of agricultural operations as presented in the Concept Note for the Agricultural Action Plan FY13–15 (May 10, 2012). Note that using this definition, it is apparent that a significant share of lending that supports agriculture is handled by sector boards other than the ARD sector board (14% in pre-crisis period and 24% in the post-crisis period). Lending for agricultural components within projects mapped to the ARD board increased by 31% between the pre-crisis period 2006–08 and the post-crisis period 2009–11 (from US$7.6 billion to US$9.9 billion), while the growth in agricultural lending by sector boards other than ARD was much higher, at 148%, albeit from a relatively small base (US$1.2 billion). These figures are in current prices, but inflation was relatively modest in the 2009–11 period.

[28] The definition utilized in the present study to estimate the volume of IFC's agricultural operations IFC operations exclude operations in forestry and non-food items (e.g. rubber), but include agricultural chemicals (e.g., fertilizers). Consequently, they differ slightly from standard presentations of figures on IFC agricultural operations.

[29] World Bank, Implementing Agriculture for Development: World Bank Group Agriculture Action Plan FY2010–12. Washington D.C., July 2009.

[30] Index-based insurance pays out pre-assigned compensation for losses resulting from weather based on measurable indicators (index) such as rainfall. Policyholders qualify for payouts as soon as the statistical indexes are triggered, without having to wait for claims to be settled or actual loss to be proven.

[31] The Syngenta Foundation for Sustainable Agriculture and UAP Insurance received US$2.4 million to help insure 20,000 farmers in Kenya over 2010–12. The International Livestock Research Institute received US$154,000 to help insure 5,000 livestock herders in northern Kenya over 2010–11. MicroEnsure received US$1.6 million to help insure 15,000 farmers in Rwanda over 2010–12.

[32] IFC, Development Outcome Tracking System, 2012.

[33] IEG, Results and Performance of the World Bank Group 2012, Vol II, August 2012, Appendix H, p. 82.

[34] IFC, Development Outcome Tracking System, 2012.

[35] Independent Evaluation Group, Growth and Productivity in Agriculture and Agribusiness, World Bank, 2011, p. 72.

[36] The December 2010 Progress Report of the GFRP stated: "In the absence of any significant incremental budget, the cost of doing crisis business was high pressure on work/life balance for involved staff and/or that staff had to be taken off "regular" program work to make time to work on GFRP project preparation".

[37] World Bank, "World Bank Group Agriculture Action Plan 2013–15", 2012. (p. 59).

# 4

# Bank Support to Social Safety Nets

## CHAPTER HIGHLIGHTS

- Governments adopted a number of measures to deal with rising food prices, including relying on distortionary policies rather than on social safety nets. A key reason is that many countries—including LICs affected by the food crisis—did not have large and well-targeted and administered safety nets in place before the crisis hit.

- The Bank's response to the global food and economic crises shows limited emphasis on assistance to populations most vulnerable to malnutrition: children under two and breastfeeding women.

- The Bank had substantive work on social safety nets for crisis response that became the basis for policy advice to mitigate the impact of the food crisis on the poor in the short term and to build resilience to future crises, although this advice was mostly based on middle-income countries.

- Countries accessing GFRP funds with limited or no safety net programs expanded in-kind transfers, especially school feeding and public works programs. Most programs were small and had limited coverage, making significant poverty alleviation impacts unlikely.

- Funding from the Rapid Social Response Program enabled work on social safety nets in the context of crisis response capacity in LICs, which may help enhance future resilience. Increasing lending in LICs that focuses on building safety net systems is a promising sign.

- The volume of analytical products, mainly nonlending technical assistance in LICs, increased considerably after the crisis period. While institutional strengthening is key for building safety nets, the reduction in ESW in the face of expanding lending to low-income and new clients risks adversely affecting the quality of the portfolio.

- Middle-income countries (MICs) continued to receive the largest share of social safety net lending. Its focus appears to be shifting to resilience building.

- Country case studies indicate that the Bank should have been engaged in social safety nets earlier.

Throughout the past decade, countries and the Bank focused social safety net support on addressing chronic poverty and human development rather than on addressing shocks such as the food price crisis. In the last few years of the decade, and largely as a result of the food and global economic crises, the focus of social safety nets shifted to addressing systemic shocks in the short-term and building resilience to better manage future crises in the medium-term.[1]

## Social Safety Net Policy Responses to the Crisis by Governments

Faced with rising food and fuel prices, governments crafted a broad range of responses, often relying on distortionary policies rather than on social safety nets. One reason these policy choices were made was that many countries did not have large, well-targeted social safety net programs in place before the food price crisis hit. A 2008 survey of IMF country desk officers covering 146 countries found that 84 countries had reduced food taxes, 29 countries had increased food subsidies, but only 39 countries had expanded their social safety nets.[2] When the food crisis hit, social safety net systems in many countries were not well prepared (Appendix H).[3] A survey of social protection staff conducted for the earlier IEG social safety net evaluation[4] found that only in 16 percent of the 65 countries studied were social safety nets considered well positioned to respond to the food and economic crises, including being able to identify and address the needs of those affected by the crises. In 40 percent of the countries, social safety nets were considered "somewhat" prepared; in 32 percent they were considered to be prepared "a little;" and in 11 percent they were considered not prepared "at all."

Another reason for the choices governments made is that some distortionary policies are easier to implement and more politically attractive in the short run, even though the measures are regressive and hard to remove.[5] Even some countries with satisfactory social safety net programs in place employed flawed policies, among them Jamaica, Mexico, and Pakistan (price subsidies or price controls, whether voluntary or enforced), Egypt and India (rice export bans), and Georgia (a one-time universal cash transfer, used instead of its Targeted Social Assistance Program).[6]

In countries with limited or no social safety net programs, governments expanded in-kind transfers, especially school feeding programs and public works programs. The expansion of school feeding programs has been a popular government response to the food crisis, particularly in Africa (Benin, Burundi, Côte d'Ivoire, Ghana, Guinea, Liberia, Lesotho, Mauritania, Mozambique, Senegal, and Sierra Leone). Unfortunately, few data are available to assess the coverage and the incidence of that expansion. A number of these programs were supported by the Bank, mainly under the GFRP. Existing public works programs were expanded to provide poor households with a source of income in countries such as Bangladesh, Cambodia, Ethiopia, Jamaica, Mexico, Nepal, and Peru.

Less commonly, some governments started new social safety net programs. For example, Liberia introduced a cash-for-work public works program and the Republic of Yemen promoted a public works program through its social fund. Indonesia reintroduced an unconditional cash transfer (UCT) program used in 2006 to remove fuel subsidies. Several countries piloted new cash transfer programs, including Afghanistan (a UCT targeted to poor families), Bangladesh (a 100-day employment guarantee scheme to provide employment to the rural poor), Democratic Republic of Congo (testing both a conditional cash and in-kind transfer program), Mozambique (public works), and Tanzania (a community-based conditional cash transfer, or CCT, program).

A startling gap in the response of governments and the Bank has been nutrition interventions for infants and mothers. Only a few countries appear to have emphasized nutrition support to this population as part of their response to the food crisis (Kyrgyz Republic, Lao PDR, Senegal, and Tajikistan, LICs, and Guatemala, Panama, and Peru, among MICs). This finding is the more startling as 22 of the 36 countries with 90 percent of the global burden of stunted growth in children, and 21 of the 32 smaller countries with more than 20 percent child stunting or underweight are among the countries "most vulnerable" or "vulnerable" to a food price crisis according to the index used by this evaluation.[7] This finding underscores the challenges the Bank and client countries face to address the operational complexities arising from the multisectoral nature of both determinants of malnutrition and nutrition interventions. It is also indicative of the low priority given to nutrition by client countries as well as the institutional barriers to cross-sector collaboration both inside the Bank and in client countries. Finally, the Bank also has traditionally had few nutrition experts on its staff.

## Bank Policy Advice on Social Safety Nets in the Crisis

The Bank had substantive analytic work on the use of social safety nets for crisis response, though there was no specific ex-ante advice on the food price crisis of 2007–08.[8] The lessons indicated that, in the short term, the causes, transmission channels, and main poverty impacts of a crisis need to be assessed at the country level as a basis for country-specific responses. The response needed to focus on protecting pro-poor social and safety net expenditures and on expanding large and effective safety net programs to operate in a "countercyclical" fashion and act as "automatic fiscal stabilizers." Safety net programs could comprise cash transfers, public works programs, and human development interventions. For the medium-term, the lessons underscored the critical need to have a safety net in place before a crisis occurs, as putting programs in place takes time. Safety net programs should be able to address the needs of the poor in normal economic times and be adaptable to address the effects of a crisis. In addition, the Bank had the outputs of an extensive Safety Net Primer program, which gathered and shared knowledge about the design and implementation of social safety net programs around the world.[9] The Human Development Network (HDN) and Social Protection Unit (SP) also provided operational guidance to staff on how to address the food crisis. Several documents on the food and economic crises appeared in 2008[10] as part of the overall institutional response. Pre-crisis advice was refined to address the specifics of the crises, even though it was based mostly on middle-income countries, which gave social safety nets a more prominent role in their poverty alleviation strategies.[11]

Rising food prices may negatively affect human development by increasing poverty, worsening child nutrition, reducing the use of health and education services, and depleting assets of the poor. As incomes fall, households switch to lower-cost cereals and less expensive sources of protein that can worsen child nutrition and result in poorer health, lower cognitive abilities, less learning, and lower lifetime earnings.[12] Children under age two,[13] pregnant and breastfeeding women, and those already suffering from malnutrition are most susceptible. Research indicates that young girls in poor families and infants born since the crisis began are most at risk of suffering irreversible damage to their physical and mental development. Evidence shows increased gender disparities in the quantity and quality of food consumed during a crisis, with mothers forgoing meals and boys getting preference over girls.

The effects of the crisis on malnutrition and schooling were identified as a threat likely to undermine years of progress on the Millennium Development Goals.[14] When the poor have to spend more on food, they have less to spend on education and health services, reducing their effective future productivity. Large numbers of children were removed from school in some locations when food prices rose, while in others parents cut back other expenses to keep them in school. It is unclear what factors help keep children in school during a food crisis (flexible schooling fee payment systems, cash transfers, or school feeding).[15] The food price crisis has eroded the savings and assets of many households leaving them with few resources to recover and manage future shocks. Therefore, continued global food price volatility is a major ongoing concern. Bank policy advice aimed to minimize the impact of crisis events on the poor and vulnerable in the short-term and help build resilience to future crises in the medium-term. In the short-term, the Bank advised scaling up the benefits or coverage of existing safety net programs, assuming that they are well targeted and well administered, rather than creating new programs. Box 4.1 shows a loose ranking of priority targeted

1. Expand benefits and coverage of existing targeted cash (or near cash) transfer programs.

2. Introduce targeted nutrition interventions for infants and pregnant women to help households use their resources most effectively to nourish their children and improve micronutrient intake.

3. Introduce in-kind food programs, including school feeding and distribution of fortified calorically dense food for children aged 0–2.

4. Expand public work programs where they exist; complement them with cash transfers.

5. Introduce fee waivers, lifeline pricing, and other forms of targeted subsidies for poor users and consumers of basic food and energy products.

6. Introduce additional measures to prevent children from dropping out of school, such as fee waivers, subsidies for school inputs, or cash transfers.

SOURCE: HDN & PREM Rising Food and Fuel...op. cit. p.1. For detailed guidance on program's choice and selection, appropriate context, advantages, disadvantages, implementation challenges, etc. see HDN Guidance...op. cit, Annex 2 "Characteristics of social safety net Interventions" and Annex 3 "Briefing Notes on Common Program Interventions."

short-term interventions in a food crisis. In the medium-term, the Bank advised helping countries build sound safety net systems so that they are better prepared for future shocks.[16]

Effective nutrition and health interventions are needed to complement safety net programs in a food price crisis in the short term. These interventions should focus on the window of opportunity from just before conception to age two. Effective interventions include nutrition education and growth promotion, targeted food supplements and micronutrients, and other primary health care interventions that reduce the risk of malnutrition.[17] Where nutrition programs do not exist, the short-term response can focus on nutrition communications campaigns, micronutrient supplementation, and the fortification of staples.[18]

Although at the time of the food crisis the Bank had a limited toolkit of social protection interventions for low-income countries and fragile states, it provided global policy guidance. In LICs, the Bank advises that safety nets focus on supplementing the income of the poorest to prevent irreversible losses of human capital or livelihoods, rather than on everyone below the poverty line[19] because of huge needs and scarce resources. In these settings, public works projects are commonly used because they not only transfer income in a self-targeted manner, but they can help build, rehabilitate, or maintain public infrastructure that will help increase productivity. However, they need to be complemented with a small cash transfer program to labor-poor households and those unable to work. Given targeting challenges in LICs (widespread poverty and lack of data), categorical targeting to particularly vulnerable groups is an option, as is community targeting. In LICs with high malnutrition, the Bank suggests using a strong nutrition program for infants and mothers that can be supported by a transfer component in coordination with the health sector. Finally, if resources exist, fee waivers for basic health and education services would support human capital formation and could be targeted to a larger population group. Implementation arrangements are a major challenge for LICs as programs will be unable to piggyback on country systems for targeting, payments, and monitoring (such as civil registries and identification documents, postal and banking systems). However, a number of LICs, especially in Africa, are piloting options for cash transfer schemes that eventually could become well-implemented safety net programs.

Regional analyses and guidance on safety net responses to the food crisis are fully consistent with the Bank's global policy advice.[20] Except in the Europe and Central Asia Region and to lesser extent Latin America and the Caribbean, Regional reports are uneven in the depth to which they analyze the distributional impacts of the food crisis. Such analysis was limited by lack of household survey data in many countries of interest (Africa, South Asia, Middle East and North Africa); in others it was limited insufficient sector knowledge (South Asia, Middle East and North Africa) or country knowledge (East Asia and the Pacific and Africa) due to lack of prior Bank engagement. The focus of regional AAA on malnutrition is limited except in Europe and Central Asia and Latin America and the Caribbean, where it is part of the recommended short-term response to the food crisis in most affected countries with high malnutrition burdens. While the Regions endorsed the Bank's global social safety net policy advice, implementation differs across regions depending in part on their countries' initial poverty conditions, affected populations, and existing safety net programs. It also depends on previous Bank engagement and sector knowledge, which was limited in some of the countries vulnerable to the food price crisis (Burkina Faso, Central African Republic, Comoros, Djibouti, Ghana, Guinea, Lao PDR, Mongolia, Nepal, and Togo. Finally, implementation is affected by IDA ceilings for LICs and for those MICs that get IDA financing (blend countries). While the global technical advice to address many of these challenges provides the programmatic options to do so, there is a gap between global advice and feasible approaches that can be implemented in LICs and fragile states.

Ex-post assessments of the social safety response to the food crisis contribute to global safety net policy guidance[21] and reinforce the necessity of having an appropriate social safety net in place *before a crisis*. The key is whether a country operates one or more high-coverage, targeted programs with sound administrative systems that can be used for short-term response (Appendix H).

The recently issued World Bank Social Protection and Labor Strategy incorporates several lessons from the food and economic crises and commits to increasing the Bank's sectoral engagement in LICs. First, recent crises, economic volatility, climate change, and natural disasters demonstrate that social safety nets are both a necessary and an effective policy tool. Second, social safety nets must be built in normal times, as the institutions and administrative capacity required cannot be built or scaled-up overnight. In normal times, social safety nets help the poor and vulnerable build resilience against idiosyncratic shocks and provide equality of opportunity. Third, to improve their crisis readiness, social safety nets must have the ability to quickly expand existing programs, identify who has been affected, and provide support without discouraging work effort or distorting markets. Fourth, prudent social safety nets are affordable in many countries. All over the world, social safety nets account for 1–2 percent of gross domestic product (GDP), even in countries with generous programs (such as Mexico and Brazil). The new strategy's operational agenda to strengthen social protection in LICs and fragile states includes, first, ensuring policy coherence across programs, donors, and government agencies; second, harmonizing and reinforcing complementarities among programs; and third, building operational subsystems that can be shared across programs, such as targeting, beneficiary registry, payment, and monitoring and evaluation systems.

## Short-Term GFRP Response and Social Safety Net Activities
SOCIAL SAFETY NETS IN THE GFRP FRAMEWORK[22]

The GFRP framework document follows to a significant extent the recommendations of the global AAA, but foresees the likely use of food as a short-term crisis response in LICs. The objective of the social protection component of the GFRP is to ensure food access and minimize the nutritional impact of the crisis. Social safety nets are recognized as the best approach in the short-term for offsetting the effect of food price increases and smoothing consumption, with minimum negative impacts on economic incentives. Direct transfers to households, especially cash, are also preferred to programs that may alter market prices and distort incentives. [23]

The GFRP framework document envisions a partnership with WFP for implementation, which is likely to have influenced the selection of social safety net instruments used under the program.[24] At the same time, it was reported that the WFP had school feeding programs in 71 of the 108 low- and lower-MICs.[25] Safety net activities under the GFRP could include "rapid response diagnostics" work for targeting and program design, "short-term financial support" to specific social safety net programs and most vulnerable populations, and "medium-term capacity building" activities to strengthen social protection systems and to build resilience in responding to future crises.[26]

## Implementation of the GFRP Social Safety Net Activities

The GFRP's safety net lending ($523 million) was a minor contributor to the expansion of the Bank's social safety net post-crisis lending ($9,229 million), but increased social safety net lending to LICs by 38 percent. The GFRP had 33 operations with social safety net activities in 27 countries (60 percent of GFRP operations).[27] Moreover, 19 of the 33 GFRP social safety net operations (58 percent) and $101 million of its commitments (or 20 percent) went to fragile states compared to 20 percent of the operations and less than 3 percent of the commitments in the regular social safety net portfolio.

The instruments most frequently used in the GFRP were public work programs and in-kind transfers, while cash transfers saw limited use (Table 4.1). This reflects the dominance of Africa in the program, where these were the existing instruments; the GFRP framework promoting work with the WFP, which runs school feeding programs in many LICs; limited country engagement and analytical work to underpin social safety net project design, compounded by time pressure to deliver quickly. Higher priority was assigned to short-term crisis response than to enhancing longer-term resilience.

Practically all GFRP safety net funding (96 percent) went to countries "most vulnerable" or "vulnerable" to the food price crisis, that is 23 out of 27 countries (see Appendix H). Africa accounted for more than half of the GFRP operations with social safety net activities and almost a third of the GFRP social safety net commitments. Fourteen African countries[28] had 18 operations. Of these countries, Comoros, Liberia, Sudan, and Togo never had a social safety net operation with the Bank before, and nine had no recent Bank-supported social safety net operations.[29] Within Africa, three countries (Ethiopia, Kenya, and Tanzania) received 63 percent of the GFRP resources for social safety net activities. Three more countries received between $10 million and $14 million (Madagascar, Senegal, and Sierra Leone); the rest received an average of $2.75 million each.

TABLE 4.1  Activities Supported in 33 GFRP Social Safety Net Operations

| Social Safety Net Instrument | Number of GFRP Operations |
|---|:---:|
| **Short-Term Responses** | |
| Cash Transfers | 5 |
| Public Work Program | 11 |
| In-Kind Transfer | 10 |
| **Medium-Term Responses** | |
| Direct Support to Government (Training and TA) | 9 |
| Targeting System | 4 |
| Payment System | 0 |
| Management and Information System | 1 |
| Monitoring and Evaluation System | 1 |
| Governance and Accountability System | 1 |

SOURCE: World Bank data.

NOTE: Activities/instruments are not mutually exclusive, so the column will not sum up to 33 GFRP social safety net operations.

Most GFRP operations with safety net activities were too small to have had an impact on reducing the effects of the food crisis on the poor. More than three-quarters of the GFRP commitments for social safety net activities (or $408 million) were allocated to only six countries: Philippines, Bangladesh, Kenya, Nepal, Ethiopia, and Tanzania (in order of commitment size). The other 21 countries received amounts ranging from $1 million (Comoros and Guinea-Bissau) to $14 million (Madagascar) and averaging $5.5 million per country, although 12 countries received $5 million or less (Appendix H).

The need for speedy crisis response appears to have driven the financing arrangements used by the safety net projects under the GFRP. First, a large proportion of the GFRP safety net resources (42 percent) was channeled through DPOs, even though there are only eight DPOs in the GFRP social safety net portfolio (Bangladesh, Burundi, Cambodia, Djibouti, Haiti, Madagascar, Philippines, and Sierra Leone). With the exception of Bangladesh and the Philippines, DPO policy matrixes lacked clear results frameworks connecting activities, indicators, and expected outputs and outcomes. The main objectives of most DPOs were to protect core spending on health, education, safety nets, and mitigate the impact of the crisis on the poor rather than institutional and policy reforms. Second, unlike the regular social safety net portfolio, the majority of the GFRP social safety net operations used additional or supplemental lending arrangements. In fact, 16 out of the 33 GFRP social safety net operations employed this modality.[30]

In-kind transfers were mainly the expansion of school feeding, which is not a first-choice instrument. Few projects incorporated measures to enhance efficacy. Most school feeding programs consisted of a snack or a meal at school (Burundi, Cambodia, Central African Republic, Djibouti, Guinea-Bissau, Haiti, Liberia, Nicaragua, Sierra Leone, and Togo). Research shows that school feeding programs have a major limitation from a nutrition perspective: they do not focus on the most vulnerable period for malnutrition and irreversible loss of human capital, which is between conception and age two.[31] From an education perspective, research shows that school feeding programs are not likely to substitute for a well-performing education program, but can enhance a system's effectiveness. However, for African LICs, the cost of school feeding programs per beneficiary is estimated to be as much as the annual per-student expenditures on education. Research also shows several complementary measures could improve the efficacy of school feeding programs, including nutrition and hygiene education, deworming, and access to micronutrients and school health. In general, the school feeding programs under the GFRP did not include these measures and/or used them to reach infants and mothers. The few exceptions include Liberia, where the school feeding programs included some take-home rations for girls and nutritional supplements and/or education for pregnant women; Togo, where the programs include training and strong participation by mothers; and Lao PDR, where primary school children received weekly micronutrient supplementation and deworming twice a year.

From a safety net perspective, research shows that school feeding programs may not be as effective as other income support programs but may be feasible in difficult contexts. They have had mixed effects on school enrollment and attendance, nutritional status, and ability to pay attention in class. In addition, the costs associated with food acquisition, transportation, storage, packaging, distribution, and preparation are higher than they are for cash, food stamps, or vouchers. Some argue that where food prices are increasing rapidly, food assistance may be more effective for the poor.[32] School feeding programs also have the potential to create distortions in food markets resulting from procurement, transport, and the distribution of food.[33] Finally, the programs are usually targeted geographically and include all children in selected schools. As a result, large-scale school feeding programs are likely to include children from less poor families (and thus have errors of inclusion). Or if they are kept small and limited to very poor areas and regions to avoid leakage, they are unable to cover the poorest children living in less poor areas (and thus have errors of exclusion). Additional research is needed on how school feeding programs compare in terms of cost-effectiveness with other social safety net transfer programs that help promote human capital investment, including targeting efficiency, impact on human capital accumulation and on household budgets, and the ability to scale-up quickly through the school network.[34]

The experience of the GFRP school feeding programs shows that the Bank worked effectively with the WFP in most of countries, although there were procurement issues in some cases such as Liberia and Burundi. Unfortunately, there are no data to assess whether the expansion of school feeding programs significantly contributed to households' income and intermediate education outcomes such as school retention and attendance. Available information shows that 60,000 Liberian children benefited from the program during the three years of the project. In Burundi, an additional 120,000 children benefited from the program in three poor regions, increasing coverage of schools by 35 percent, and daily meals delivered in schools by 68 percent. In Nicaragua, an additional 300,000 children received school lunch during 1.5 months of the school year under GFRP financing. Other school feeding programs appear to be very small for significant coverage of poor school children and poverty impact. For example, 14,000 children benefitted in Guinea-Bissau and 3,700 children with disabilities were assisted in Sierra Leone. While most programs declare to have targeted "poor areas," there is no data to assess their actual incidence.

GFRP projects also sought to quickly scale-up existing public works programs, a suitable safety net instrument proven feasible in LICs and fragile states. GFRP projects with such activities (either cash or food-for-work) were implemented in Bangladesh, Comoros, Ethiopia, Guinea, Guinea-Bissau, Haiti, Madagascar, Nepal, Sierra Leone, Sudan, West Bank and Gaza, and the Republic of Yemen. They financed the continuation or expansion of existing programs to more people or food insecure areas under a variety of institutional setups. In addition to providing poor workers with a source of income, these programs create, rehabilitate, and maintain public infrastructure—examples include rural roads, river embankments, irrigation canals (Bangladesh); irrigation and drinking water infrastructure, small bridges, community buildings (Nepal); basic social infrastructure such as schools, health facilities and rural roads (the Republic of Yemen); rural roads rehabilitation and environmental rehabilitation (Sierra Leone). Common issues with public works programs include inadequate wage rate (too high for the work effort required, as in Madagascar or many rates or payment forms, as in Bangladesh); financing of "private goods" as opposed to public goods (the Republic of Yemen); lack of maintenance of constructed/rehabilitated infrastructure (the Republic of Yemen); insufficient participation of women (Madagascar and the Republic of Yemen appear to have been more successful ensuring women's participation); and governance issues (Bangladesh which has several public works programs).

The largest public works program supported by the GFRP (additional finance) was part of Ethiopia's flagship Productive Safety Net Program (PSNP), which combines cash transfers and public works. The PSNP's impact evaluation shows that after five years participating in the PSNP, households enhance their food security by more than 1.5 months during a 12-month period, increase livestock and the value of their productive assets, and sell fewer assets in distress. Nepal's public works program reports that 168,000 beneficiaries improved their households' food security by about two months during a 12-month period. For other public works programs, available information only indicates the number of people benefitting from the program or the number of jobs created. For example, Comoros reports 1,100 beneficiaries; South Sudan 24,000; Guinea mentions 5,300 unskilled jobs; and Madagascar reports 92,000 beneficiaries.

Safety net activities under the GFRP seem to have provided an "entry point" for policy dialogue on social safety net options for crisis response in the longer-term. A few of the projects included institutional development activities. Regional staff used the opportunity to get traction on country social protection and social safety net diagnostics (major issues, existing institutions and programs, and financing issues), and start the policy dialogue necessary for the definition of a longer-term strategy to enhance resilience with governments and donors. To this end, support through catalytic grants from the Rapid Social Response (RSR) program for social safety net systems building in LICs has provided critical resources for staff to engage with countries in the analysis of longer-term issues, enable country experimentation and piloting, and finance technical assistance for the preparation of safety net operations focused on resilience building. Of the 27 GFRP countries with social safety net activities, 18 have received support from the RSR Program. According to the Social Protection Unit, as of January 31, 2012, $25.5 million in RSR resources in 26 countries were accompanied by $1.33 billion of Bank loans and grants (approved and in the pipeline).[35]

Many countries with GFRP activities have follow-up operations focusing squarely on safety nets, a promising sign for the resilience building agenda. According to the IEG evaluation database, at the end of FY2011, 20 of the 33 GFRP operations with social safety net activities had already closed and 13 operations (in 12 countries) remained active.[36] By the end of FY2011, the Bank had approved 17 new regular social safety net operations in 13 of the 27 GFRP economies that had social safety net activities: Bangladesh, Comoros, Ethiopia, Kenya,

Liberia, Moldova, Nicaragua, Philippines, Sierra Leone, Tajikistan, Tanzania, West Bank and Gaza, and the Republic of Yemen. Specific examples of resilience building include Djibouti with a pilot of an integrated social safety net including public work programs, social transfers, and nutrition; Tajikistan which is already implementing a project to enhance government's capacity to plan, monitor, and manage social assistance to the poor; and Liberia, which is expanding small public works programs based on positive evaluations of the programs' targeting and impact on household income with a project supported by IDA under the Crisis Response Window and other donors. Perhaps this indicates a welcome shift in the Bank's social safety net engagement pattern in LICs from ad hoc and opportunistic to a continuous relationship and joint work over time, which appears essential to build both the social safety net programs and the basic systems capable of responding to systemic shocks.

Although lack of large and well-targeted and administered safety net programs constrained the adoption of global advice, implementation highlights several issues with project design choices:

- **Disconnect between GFRP framework and policy advice and limited operational focus on nutrition.** Few GFRP safety net projects included actions to provide nutrition support to children under two and pregnant and breastfeeding women, even though this was a clearly stated recommendation in the global AAA, a priority for the GFRP, and the majority of GFRP countries are among those with the highest malnutrition burdens in the world. Only Kyrgyz Republic, Lao PDR (a pilot), Liberia (small sub-component), Moldova, Nepal, Tajikistan, Sierra Leone and Senegal focused on infant and maternal nutrition. The RSR Program is including significant support to nutrition activities,[37] with Africa receiving particular emphasis (Benin, Ethiopia, Gambia, Madagascar, and Malawi, plus several regional and subregional initiatives). Other countries include India, Tajikistan, the Republic of Yemen, the IDA eligible countries in LCR, all priorities from the standpoint of child malnutrition.[38] The limited operational emphasis on child nutrition during the post-crisis period can also be seen in the overall Bank response to the global economic crisis, in spite of the Bank's efforts to scale up nutrition activities (including hiring of new staff, RSR support to nutrition activities in Africa and South Asia).

- **Targeting, coverage, and incidence issues.** Identification of the key groups to be assisted with a social safety net is a challenge both in stable and crisis times as most developing countries lack appropriate household data. Scarcity of data needed for household targeting as well as adequate governance arrangements to manage targeting, intake, exit, and appeals are more acute for most of the GFRP countries. Very few of the countries studied for this evaluation provided an assessment of the impact of the crisis on the poor (Bangladesh, Nepal, and Nicaragua were exceptions). More GFRP projects with social safety net activities describe mechanisms to be used to select beneficiaries, mostly using a combination of geographic and then community targeting, a practical approach that can produce adequate targeting outcomes in data-constrained environments. Many GFRP projects do not specify expected or actual social safety net program coverage and incidence to assess the likely contribution of the project to the population in need of assistance and whether they reached the intended target populations. Instead, they report indicators such as the numbers of children to receive food in school, or number of hospital patients to be fed (such as in Burundi, Liberia, and Sierra Leone), and mention that project activities were targeted to food-insecure areas.

- **Questions on efficacy and effectiveness of programs.** Some project designs were missing key elements that determine effectiveness. For example, for public works programs to meet social safety net objectives they need to have a clear targeting method to select locations, low wages to have self-selection of poorer workers, high labor intensity and the use of unskilled labor, a portfolio of community level investments (infrastructure, environment, community services), meaningful duration (number of workdays per worker), exit rules, and a good management information system to assess a program's effectiveness as well as to monitor transparency of operations. These lessons are not often applied, ostensibly due to country political economy issues, including existing practices by other donors (using higher wage levels, for example) that affect government choices (as occurred in Madagascar). School feeding programs are another example where measures to enhance efficacy (nutrition and hygiene education, and access to micronutrients and school health) were not generally adopted.

- **Limited background knowledge and time pressures.** Analytical work to underpin GFRP social safety net lending was extremely limited. This reflects the lack of previous Bank engagement in many GFRP countries, and probably constrained project design choices, including selection of social safety net interventions, targets, and indicators. The country studies conducted for this evaluation indicate that in most GFRP countries the Bank did not conduct a rapid country safety net diagnosis as the basis for project design. In some cases, it assessed crisis impact on poverty (in Bangladesh, for example). Instead, the Bank used previous economic and sector work (in Bangladesh, Ethiopia, Kenya, Kyrgyz Republic, and Madagascar), or assessments by other donors (as in Ethiopia, Kenya, Kyrgyz Republic, and Nepal,). However, with the most notable exceptions of the Philippines and Ethiopia, the majority of the country studies show that the analytical bases for project design choices were not clearly spelled out. Speed of response seems to have been the overriding criterion for selecting objectives, specific interventions, targets, and indicators (Burundi, Djibouti, and Sierra Leone, for example). This raises questions about how relevant and appropriate the social safety net interventions selected for support under some GFRP projects were relative to their objectives, and the quality of design of their results frameworks.

## Longer-Term Response
ANALYTIC AND ADVISORY ACTIVITIES

AAA activities increased over FY2009–11 but only on nonlending technical assistance (NLTA),[39] and mainly in LICs. However, the majority of the social safety net AAA resources continued to go to MICs (Table 4.2). The number of social safety net ESW activities declined from 86 in FY2006–08 to 78 in FY2009–11, while the number of NLTA activities rose from 36 to 89 during the same period. Commitments in ESW remained practically unchanged while NLTA commitments almost doubled, mainly in Africa, Europe and Central Asia, and Latin America and the Caribbean (Table 4.3). Still, the majority of the social safety net AAA resources continued to go to MICs, which received over 64 percent of the AAA resources in FY2009–11 ($10.1 million), with LICs receiving just 14 percent ($2.2 million).

TABLE 4.2  Social Safety Net Analytic and Advisory Product Lines FY2006–11

| Product Line | Number of AAAs | | | Commitments (US$ Millions) | | |
|---|---|---|---|---|---|---|
| | Pre-Crisis (FY2006–08) | Post-Crisis (FY2009–11) | Percentage Change | Pre-Crisis (FY2006–08) | Post-Crisis (FY2009–11) | Percentage Change |
| ESW (Total) | 86 | 78 | −9 | 7.8 | 8 | 3 |
| LICs only | 16 | 13 | | 1.5 | 1.2 | |
| Technical Assistance (Total) | 36 | 89 | 147 | 2.6 | 7.7 | 195 |
| LICs Only | 1 | 14 | | 0.1 | 1.1 | |
| TOTAL | 122 | 167 | 37 | 10.4 | 15.7 | 51 |

SOURCE: World Bank data.
NOTE: AAA = analytical and advisory activities, ESW = economic and sector work, LICs = low-income countries.

The large increase in social safety net NLTA activities after the crisis, accompanied by a decline in social safety net ESW may indicate that the Bank is shifting its country engagement modality, but it also suggests that the analytical underpinnings for its operations may be suffering: lower ESW while significantly increasing lending may signal a lack of in-depth social safety net analytical and diagnostics work for policy advice, project design, and implementation, particularly in LICs where social protection and safety net issues have been largely absent from the country programs (South Asia, Africa, and the Middle East and North Africa). NLTA activities have a less standardized and established quality assurance process than ESW.

Funding through the RSR program financed by a multidonor trust fund and the President's Office and managed by the Social Protection Anchor complemented the scarce regional resources and enabled the increase of social safety net AAA on crisis response in LICs. Over 60 percent of RSR resources ($35 million) have been committed to 44 activities or projects to build social safety net and social protection systems in 34 eligible IDA countries so that they

TABLE 4.3 Social Safety Net Number of Analytic and Advisory Activities by Region FY06–11

| Region | FY2006–08 | | | FY2009–11 | | |
|---|---|---|---|---|---|---|
| | ESW | Technical Assistance | Total | ESW | Technical Assistance | Total |
| AFR | 14 | 7 | 21 | 17 | 14 | 31 |
| EAP | 10 | 8 | 18 | 10 | 9 | 19 |
| ECA | 26 | 2 | 28 | 25 | 26 | 51 |
| LCR | 13 | 5 | 18 | 11 | 13 | 24 |
| MNA | 8 | 14 | 22 | 3 | 13 | 16 |
| SAR | 12 | | 12 | 9 | 8 | 17 |
| Regional Studies | 3 | | 3 | 3 | 6 | 9 |
| TOTAL | 86 | 36 | 122 | 78 | 89 | 167 |

SOURCE: World Bank data.
NOTE: AFR = Africa, EAP = East Asia and the Pacific, ECA = Europe and Central Asia, LCR = Latin America and the Caribbean, MNA = Middle East and North Africa, SAR = South Asia.

are better prepared to respond to crises, natural disasters, or post-conflict situations. Activities supported by the RSR include assessments of the impact of the crisis on the poor and their human capital; analyses of existing formal and informal social safety net programs; technical assistance and capacity-building activities such as the development and start-up of targeting, a beneficiary registry and identification methods and systems, improvement of beneficiary payment systems, and development of capacity for M&E of social safety net programs; feasibility assessments of different types of programs, including cash transfers, public works programs, nutrition interventions, and others.

## POST-CRISIS LENDING PATTERNS IN THE SOCIAL SAFETY NET PORTFOLIO[40]

Pre-crisis, the Bank's social safety net portfolio concentrated on MICs. As shown in IEG's evaluation on social safety nets, between 2000 and 2010, 62 percent of Bank projects supporting social safety net activities and 79 percent of safety net lending went to MICs.[41] There are four reasons for this pattern.[42] First, social protection, in general, and social safety nets, in particular, are more likely to be an integral part of MICs' poverty reduction agendas. In LICs, many other priorities compete for scarce resources, and social safety nets are commonly considered unproductive "handouts" and that "everybody is poor."[43] Second, MICs have greater borrowing and spending capacity than LICs, which also have lower institutional

TABLE 4.4  Regular Social Safety Net Lending Pre- and Post-Crisis by Country Income Level

| Country Income Level | Social Safety Net Regular FY2006–11 | | Social Safety Net Regular FY2009–11 | |
|---|---|---|---|---|
| | Number of Ops | Amount (US$ Millions) | Number of Ops | Amount (US$ Millions) |
| HIC | | | 2 | 253.6 |
| LIC | 17 | 297.5 | 21 | 776.5 |
| LMIC | 23 | 543 | 44 | 1,724.4 |
| UMIC | 21 | 535.3 | 39 | 6,474.3 |
| TOTAL | 61 | 1,375.8 | 106 | 9,228.8 |

SOURCE: World Bank data.
NOTE: HIC = high-income country, LIC = low-income country, LMIC = lower-middle-income country, UMIC = upper-middle-income country.

and implementation capacity and tight ceilings on IDA funding. Third, in LICs, donor grant financing is available for humanitarian assistance and relief support, which is often combined with social safety net programs. Fourth, the Bank has had a sustained pattern of social safety net engagement in MICs, including a progressive shift from support to specific social safety net programs to support social safety net and social protection systems' reform and the development of broader social protection systems capable of addressing a variety of risks, especially after the global economic crisis. This has been the case in Brazil, Chile, Colombia, Mexico, and Turkey, among others.

In LICs, the pre-crisis social safety net portfolio concentrated on program-specific support, short-term emergency response, and piloting new interventions. social safety net projects were distributed more thinly over more countries, with an emphasis on emergency response and specific interventions (such as social investment or action funds, SIF or SAF) or vulnerable groups (such as orphans and vulnerable children or demobilized soldiers). Pre-crisis social safety net lending to LICs included a large number of operations in fragile states. According to the IEG social safety net evaluation, the Bank supported 26 of the 33 fragile states with one or more projects between 2000 and 2010.[44] The Bank's involvement has used social safety net activities as a tool in post-conflict recovery.[45] At the same time, experience in

TABLE 4.5 Regular Social Safety Net Lending Pre- and Post-Crisis by Region

| Region | FY2006–08 | | FY2009–11 | | Percentage Change | |
|---|---|---|---|---|---|---|
| | Number of Ops | Amount | Number of Ops | Amount | Number of Ops | Amount |
| AFR | 8 | 185.1 | 24 | 803.9 | 200 | 334 |
| EAP | 4 | 122.9 | 9 | 704.7 | 125 | 473 |
| ECA | 11 | 166.3 | 26 | 1,903.2 | 136 | 1,044 |
| LCR | 21 | 499.4 | 28 | 5,112.3 | 33 | 924 |
| MNA | 8 | 53.4 | 9 | 152.7 | 13 | 186 |
| SAR | 9 | 348.7 | 10 | 552 | 11 | 58 |
| TOTAL | 61 | 1,375.8 | 106 | 9,228.8 | 74 | 571 |

SOURCE: World Bank data.
NOTE: AFR = Africa, EAP = East Asia and the Pacific, ECA = Europe and Central Asia, LCR = Latin America and the Caribbean, MNA = Middle East and North Africa, SAR = South Asia.

some LICs—such as Ethiopia and Moldova, which have had a long-term social safety net engagement with the Bank—demonstrate that it is possible for LICs to have sound social safety net strategies and systems, and prudent and affordable programs (Table 4.4).

After 2008, social safety net lending expanded exponentially, with MICs absorbing most of this expansion, suggesting that without special programs—such as the GFRP and the Special IDA Crisis Response Window—the Bank's social safety net response to the food crisis in LICs would have been more limited (Table 4.5). In all MICs, the number of operations with social safety net activities increased by 75 percent and commitments rose almost sixfold (from $1,078 million pre-crisis to $8,199 million post-crisis). Continued dominance of MICs in social safety net lending is not only because of their greater absorptive capacity compared to LICs, but for two other reasons as well. First, the food and economic crises affected different countries in different ways and at different stages, requiring an escalation of social safety nets to respond to aggregate shocks. This was compounded by the severity of the financial crisis in MICs, especially in ECA and LCR. (Mexico alone accounts for $2.8 billion of post-2008 social safety net lending.) Second, a number of the MICs affected by the economic crisis had long-term continuous social safety net engagement with the Bank and increased demand for the expansion of social safety net programs and system's strengthening.[46]

Post-crisis social safety net lending to LICs was greatly facilitated by the Rapid Social Response Program. The RSR was formally created by the SP Sector Board in December 2009,[47] but it initiated its operation earlier with resources from the President's Contingency Fund ($920,000).[48] The RSR Program attracted trust fund support for a total value of $61.7 million.[49] The RSR supported Bank-executed AAA (ESW, technical assistance, capacity building), direct grants for specific social safety net systems, piloting and scaling-up programs, and knowledge management activities (this last activity included both IDA and IBRD countries). The RSR received $2.8 million from the UK's Catalytic Fund, $58.5 million from a multidonor trust fund (Russia and Norway), and $3.92 million from the President's Central Contingency Fund). Almost all the RSR funds (95 percent) have been used for country or region-specific technical assistance and pilot projects, with Sub-Saharan Africa taking in almost 50 percent of the funding. According to the IEG social safety net evaluation, the RSR has been responsible for the expansion of the Bank's social safety net activities in LICs.[50]

Driven by the need to respond quickly to the crisis events, the use of DPOs to support social safety nets spiked in FY2009–11, increasing from a quarter of regular social safety net operations (16) in the pre-crisis period to almost half (51) in the post-crisis period. This increase took place mostly in MICs (where the number of DPOs increased from 11 to 47 in the same periods) and fragile states (where the number of DPOs rose from 2 to 8). The number of completed DPOs with social safety net components that have been reviewed

TABLE 4.6 Social Safety Net Instruments Pre- and Post-Crisis Periods by Country Income Level

| Type of Social Safety Net Intervention/ Social Safety Net Instruments[a] | Number of Operations with Instrument | | | | | |
|---|---|---|---|---|---|---|
| | LICs | | | MICs | | |
| | FY2006–08 | FY2009–11 | Percentage Change | FY2006–08 | FY2009–11 | Percentage Change |
| Conditional Cash Transfer (CCT) | 2 | 1 | −50 | 14 | 30 | 114 |
| Unconditional Cash Transfer (UCT) | 3 | 9 | 200 | 10 | 41 | 310 |
| Public Works Program (PWP) | 3 | 12 | 300 | 6 | 20 | 233 |
| In-Kind Transfer | 4 | 3 | −25 | 7 | 8 | 14 |
| Health and Education Subsidies | 7 | 2 | −71 | 7 | 18 | 157 |
| Water, Energy and Other Subsidies | 1 | 2 | 100 | 9 | 14 | 56 |
| TOTAL NUMBER OF PROJECTS | 17 | 21 | | 44 | 83 | |

SOURCE: World Bank data.
NOTE: LICs = low-income countries, MICs = middle-income countries.
a. These instruments are not mutually exclusive and therefore do not add to total number of projects.

by IEG is too small to draw conclusions about whether such operations achieved their development objectives.[51] Unlike the GFRP social safety net operations, the regular social safety net portfolio shows only a slight increase in the number of additional and supplemental finance operations in FY2009–11 (from 12 to 16, or 17 percent and 13.4 percent of the portfolio, respectively).

Safety net instruments[52] used for crisis response varied by country income level, and region (Table 4.6). MICs' social safety net response to the global economic crisis focused on cash transfers (conditional and unconditional) followed by public works programs, even though the use of targeted health and education subsidies also grew considerably. These findings reflect an expanded use of pre-existing social safety net instruments in the countries. They also reflect the extensive use of public works programs as a response to the impact of financial crisis and associated increase in unemployment, under-employment, and reduction in remittances, especially in Europe and Central Asia and Latin America and the Caribbean. Finally, the findings reflect adherence to global AAA advice on cash or near cash transfers as the preferred instruments for crisis response wherever possible. However, in an inflationary crisis, "cash" may not be the most effective tool unless the value of the transfer keeps pace with inflation. For example, during the food crisis, in Ethiopia's PSNP (which transfers both cash and food), the cash portion of the program significantly eroded in value, creating a preference among beneficiaries for a combination of cash and food not related to the pros and cons of food versus cash but, rather, to the value of the transfer.[53]

The most used social safety net instruments in LICs were public works programs and UCTs. Public works programs were also used for crisis response because they were already in place and part of the few existing social safety net programs that could be scaled-up in most crisis-affected countries; they are more politically acceptable than pure transfers; and they provide an additional income-earning option for poor households. The use of in-kind transfers actually declined as an social safety net response instrument in the regular portfolio, which contrasts with findings under the GFRP projects. In addition, very few social safety net projects in LICs (5 percent) supported CCTs, reflecting the countries' low capacity to monitor compliance with the conditions typically associated with this type of program.

The Sector Board managing social safety net operations also seems to make a difference for the type of social safety net program used for crisis response; less than half of the total social safety net portfolio is under the oversight of the SP Sector Board. The other Sector Boards managing social safety net operations were Economic Policy and Agriculture and Rural Development. Post-crisis operations overseen by the SP Sector Board (49 percent or regular portfolio) show increased use of public works programs, targeted health and education subsidies, and UCTs. PREM Sector Boards (mainly Economic Policy) managed a third of the post-crisis social safety nets. Their operations almost exclusively used UCTs and CCTs.

Like GFRP operations, regular social safety net operations also show limited emphasis on nutrition interventions in the post-crisis period. The number of regular social safety net projects with specific nutrition activities remained low (actually declining from 9 to 7 projects in FY2006–08 and FY2009–11) while commitments to nutrition activities increased by a meager 30 percent (from $130 million in the pre-crisis period to $170 million in the post-crisis period).[54] Nevertheless, two of the six countries that had nutrition-specific activities in their social safety net operations (Guatemala and Peru) are part of the 36 countries that account for 90 percent of the global burden of malnutrition.[55] Only Europe and Central Asia and Latin America and the Caribbean stressed the importance of nutrition interventions as a key part of the response to the crisis events, particularly in those countries with high malnutrition burdens and in the Andean and Central American countries in Latin America and the Caribbean as part of regular social safety net lending (most notably Guatemala, Panama, and Peru).

The limited emphasis in Bank lending on malnutrition does not appear to have been addressed by other sectors in the post-crisis period. A separate but cursory review of the non-social safety net nutrition portfolio[56] in the pre- and post-crisis period shows fewer nutrition projects (going from 13 in the pre-crisis period to eight in the post-crisis period) and a doubling of commitments for nutrition (increasing from $107 million pre-crisis to $221 million post-crisis).[57] In the Africa region, both the number of projects and commitments for non-social safety net nutrition activities declined (from seven to three projects and $49 million to $26 million in the pre- and post-crisis periods, respectively[58]), while South Asia began new operations in the post-crisis period (two projects and $85 million in nutrition commitments[59]). Latin America and the Caribbean also reduced its operations (from four to two) and focused on Nicaragua and Peru, the latter an upper-middle-income country with a high burden of chronic malnutrition, particularly among the indigenous populations. Following global AAA advice, the bulk of the pure nutrition portfolio focuses on children from conception to age two, their mothers, and pregnant women with maternal and child health interventions, micronutrients, and promotion of behavioral change through nutrition education and growth promotion. HDN network reports that new and pipeline nutrition commitments are increasing for FY2013 and FY2014 compared to previous fiscal years. The bulk of new lending is occurring in South and East Asia, a welcome development given that these regions have the highest global burden of malnutrition.

Efforts to enhance resilience to crisis in regular social safety net emphasized MICs, including support to the basic building blocks for equitable, efficient, and transparent social safety net operations, such as targeting, beneficiary identification and registration systems, beneficiary payment systems, management information systems, and M&E systems.[60] These building blocks and their institutions can be used to implement a particular social safety net program or across a range of social safety net programs.[61] Regular social safety net projects including institutional development objectives and activities in MICs increased from 59 percent in the pre-crisis period to 65 percent post-crisis, signaling that both the countries and the Bank are increasing their attention to the structures and institutions needed to deliver targeted assistance to poor households. Main areas of support included administration improvements, M&E capacity building, targeting systems, and beneficiary registries. In LICs, the share of regular social safety net projects with institutional development objectives and activities somewhat declined from 47 percent in the pre-crisis period to 43 percent in the post-crisis period, indicating the highest priority assigned to short-term crisis response lending, especially in Africa and East Asia and the Pacific.[62] Given that the RSR Program established to support social safety net basic systems building in LICs in the post-crisis period began to provide catalytic resources for social safety net institutional development activities in 42 LICs in 2009,[63] it can be expected that newer social safety net projects in LICs will show increased attention to building resilience.

Lack of country readiness to cope with the impact of the global food crisis on the poor underscores the relevance of the social safety nets "resilience" agenda, and requires the Bank's long-term and sustained engagement, especially in LICs. Improving social safety net country readiness in LICs requires addressing three challenges: financing arrangements, sustainable institutional capacity, and donor alignment.

- A well-targeted and prudent social safety net need not be unduly costly. Total spending on social safety nets has been around 1 to 2 percent of GDP for many developing countries. Moreover, many lower-middle-income countries and some LICs spend their own (and donor) resources in poorly targeted, inefficient, and ineffective programs and subsidies that, if reallocated, could suffice for a prudent and well-targeted social safety net.

- The central issue regarding country capacity is how to strengthen it so that systems and institutions are capable of designing and delivering a basic social safety net that can be expanded during a crisis. LICs and fragile states have infrastructure constraints and overall (non-sector specific) governance weaknesses and vulnerabilities that limit the selection of implementation arrangements for social safety net programs. The experience of a few LICs that have started to develop social protection strategies that include the establishment or strengthening of key building blocks for an social safety net, such as Ethiopia, Kenya, Rwanda, and Liberia, could provide useful lessons for other LICs with similar constraints.

- Donor alignment is essential for LICs to have an effective social safety net capable of responding to a crisis with sufficient scale. Donor support must be aligned and coordinated around specific objectives, priorities, interventions, and processes. The experience of

Ethiopia—a notable exception—illustrates what should be the aim for social safety nets in LICs: transition from ad-hoc fragmented emergency relief programs to an social safety net that is owned, led, and organized by the government and supported by several donors.

## Effectiveness and Sustainability of the Social Protection Portfolio

While the majority of social protection projects continued to perform well relative to the Bank's average, the share of completed social protection projects rated moderately satisfactory or better by IEG declined in FY2009–11. Before the crisis period, the share of completed social protection projects rated moderately satisfactory or better by IEG was well above the Bank average. The percentage of completed social protection operations with moderately satisfactory development outcomes or better declined from 83 percent in the pre-crisis to 72 percent in the post-crisis period (relevant tables are in Appendix H).

Social safety nets portfolio performance has deteriorated slightly. Two caveats must be kept in mind regarding the performance of social safety nets projects. First, even though social safety net projects are a major share of the social protection portfolio post-crisis, the data used for this analysis cover the whole social protection portfolio and include other types of projects,[64] and a majority of projects not managed by the Social Protection Sector Board. Second, the declines in social protection ratings were not shown to be statistically significant. In any case, it is of concern that all social protection indicators (except for quality at entry of DPOs) are lower in the post-crisis period. It suggests that fewer completed social protection projects achieved their development outcomes and may indicate quality issues with the ongoing portfolio as well.[65]

The social protection portfolio performed less well in upper-middle-income countries while performance in LICs and fragile states appears not to have changed much. Exploring the specific reasons for such an outcome requires careful assessment, especially if it relates to the shift in emphasis toward systems change and institution building in the regular social protection portfolio. It is widely acknowledged that the Bank has challenges with policy and institutional development where it tends to have unrealistic expectations of what can be achieved in short periods. For projects dealing with complex systems change or social protection reform, the Bank could have set goals too ambitious to be achieved by single projects, created issues on setting objectives, indicators, and targets, and further complicated the development of project result frameworks, a shortcoming already identified by the IEG social safety net evaluation.

The reasons for the slight deterioration of the social protection portfolio are diverse. The small number of completed social protection operations reviewed by IEG in the 2012 Results and Performance report limits options to identify the factors behind declining social protection projects' outcome ratings. Possible explanations for the decline include deterioration of country environment and circumstances, poor design of operations including GFRP lending with social safety net activities, and weakening of the Bank's performance as reflected by both project quality at entry and quality of supervision. The social protection lending portfolio shows deterioration in both these indicators in the post-crisis period: quality at entry dropped by 9 percent and quality of supervision dropped by 3 percent.

Several factors internal to the Bank may explain these findings. First, sectoral expertise may have been not been sufficient to manage the fivefold portfolio increase in FY2009–11. In fact, the social protection sector had its lowest staffing levels at the height of the crisis (116 GF+ staff in FY2008 and 127 in FY2009), when a large part of the lending increase took place. Second, per-task budgets declined. Given that the Bank maintained its administrative budget fixed at $1.6 billion per year while dramatically increasing both the number of operations and the volume of social protection lending, task budgets for social protection projects likely declined.[66] Stagnant or lower task budgets compounded with few SP staff at the peak of the crisis lending work had to result in an increased workload for staff. The same staff are responsible for the AAA underpinning projects, for the preparation of new operations, and for the supervision of the ongoing portfolio, thus affecting project outcomes. The combination of these factors may be at play in the ongoing portfolio as well, which merits prompt corrective action by Bank management.

## Lessons from the Bank's Social Protection Response

All countries need social safety nets in place before a crisis hits, most low-income countries were not prepared and the Bank response to the global food crisis was restricted by pre-existing social safety net programs and systems. The Bank focused its lending, analytical, and capacity building support for social safety nets significantly more on middle-income countries than low-income countries throughout the decade. There have been efforts to upgrade social safety nets in low-income countries since the food price and global economic crises of 2007–08, mainly supported by the centrally funded RSR and by special funding provided by GFRP and IDA Crisis Response Window. While there is no hard evidence yet, indications are that many countries, including Kenya, Liberia, and the Republic of Yemen among others, have stepped up efforts to improve their social safety net readiness, aided by Bank support.

In-kind programs, especially school feeding and public work programs, can be practical and politically acceptable vehicles for social assistance in countries without more sophisticated systems, as long as their limitations are recognized and longer-term options developed. In some contexts, well-designed school feeding programs can be targeted moderately accurately, though rarely so effectively as the most progressive of cash transfers. In the poorest countries, which was the case most of GFRP countries, where school enrollment is low, school feeding may not reach the poorest people, but in these settings alternative safety net options are often quite limited, and geographically targeted expansion of school feeding may still provide the only option for rapid scale-up safety nets.[67] Public work programs were also used to quickly scale up existing programs. In addition to providing a source of income to poor workers, these programs create, rehabilitate, and maintain public infrastructure, which contributes to increase productivity. However, public work programs rarely achieve a very large scale and require significant investment resources.

Capacity-building investment in early warning systems can be an important component of subsequent lending operations designed to support the building of more robust and permanent systems. The inability of governments to address emergencies due to more than the surprise factor, it is also because institutions are unprepared or financial resources are lacking. A finding of the evaluation is that the absence of sound policies and competent institutions to implement them is more often a binding constraint for speedy action than financial resources. For example in Tajikistan, following the successful implementation of a social protection operation funded by the GFRP, policy analysis and an investment operation were used to make major improvements to the country's capacity to provide social protection during emergencies.

Going forward, improving low-income countries social safety net crisis preparedness requires that the Bank maintain special efforts (financing and internal incentives) and continued engagement at the country level. Major challenges for building social safety net programs in low-income countries include very limited institutional capacity, inadequate funding and financing arrangements, and donor fragmentation.

Funding from programs such as the RSR initiative can support social safety nets work on crisis response capacity in low-income countries, which may enhance future resilience. Confirming prior findings in the IEG Safety Nets Evaluation, this evaluation found that the Rapid Social Response supported Bank-executed AAA, direct grants for specific social safety net systems, piloting and scaling up programs, and knowledge management activities in over 40 low-income countries. Many of them such as Bangladesh, Kenya, Liberia, Nepal, Nicaragua, Tajikistan, and Tanzania have started to upgrade their social safety net systems with contributions from Rapid Social Response and GFRP.

Economic and sector work on social protection in low-income countries continues to be a priority. Analytic work to underpin social safety net lending within GFRP program was extremely limited, which probably constrained project design choices, including selection of social safety net interventions, targets, and indicators.

## Endnotes

[1] HDN, Guidance for Responses from the HD Sector to Rising Food and Fuel prices, August 2008.

[2] HDN and PREM Networks Rising Food and Fuel Prices: Addressing the Risks to Human Capital, October 13, 2008, p.11.

[3] This classification of countries follows closely HDN, Grosh M. et al. Assessing Safety net Readiness in Response to Food Price Volatility. Social Protection Discussion Paper 1118, September 2011.

[4] IEG Evaluation of World Bank Support to Social Safety Nets 2000–2010, Appendix VII.

[5] HDN Guidance for Responses from the HD Sector to Rising food and Fuel prices, World Bank, August 2008 p.11.

[6] ECA, Rising Food and Energy prices in Europe and Central Asia. World Bank. 2011, p.34.

[7] See, HDN, Horton S. et al., Scaling Up Nutrition. What will it cost? World Bank, Human Development Network, Washington D.C., 2010, Chapter 2, p.6. Country vulnerability: Xue's file 87 LICs and MICs (email of 08-18-12).

[8] A comprehensive summary in Staff from WB, IMF, ADB, IDB, Social Safety Nets in Response to Crisis: Lessons and Guidelines from Asia and Latin America. Submitted to the APEC Finance Ministers, February, 2001.

[9] This work was led by the social protection anchor in partnership with the regional vice presidencies since the mid-1990s. For the full suite of products comprising the Bank's knowledge on SSNs see www.worldbank.org/safetynets.

[10] Main reports include Grosh M. et al., For Protection and…op. cit.; HDN Guidance for…op. cit.; and HDN and PREM Networks Rising Food…op. cit.

[11] This guidance included: (a) additional AAA reports and technical guidance notes on the process to carry out diagnostics of the impact of the food crisis on the poor and vulnerable, rapid review of existing programs, and review SP and SSN expenditures; (b) the creation of the Rapid Social Response Program (RSR) in FY09 to help low-income countries build SP systems so that they are ready to protect and invest in their populations in future crises; (c) tools and software for social protection analyses at the country level (see HDN Guidance…op. cit. Box 3 page 10. These tools encompass guidance and a toolkit on public expenditures analysis of SP, and software (ADePT SP) to produce rapid diagnostics of poverty, labor market, gender and SP based on STATA/SPSS routines that can be used to calculate coverage indicators, benefits level, and incidence of SP programs); and (d) in partnership with the regions, active support sharing global knowledge on SSN responses to the crises through training, South-South knowledge sharing, publications, and toolkits.

[12] HDN & PREM, Rising…op. cit. p.4.

[13] The period between conception and 24 months is the most critical time for children's growth and development. The damage caused by early malnutrition is irreversible by age 3. HDN Guidance…op. cit., p.18.

[14] HDN & PREM, Rising… op. cit. p.1. The WB Global Monitoring Report (GMR) 2012. Draft. Chapter 2, "Nutrition Needs More Attention," pp. 1–29 underscores HDN's assessment, elaborates further child nutrition issues, and presents recent research findings.

[15] Compton et al. Impact...op. cit. p.57.

[16] HDN & PREM Rising Food and Fuel...op. cit. p.1.

[17] Such as immunizations, oral rehydration therapy, protection against malaria, etc. HDN Guidance op. cit. p. 18. HDN & PREM. Rising...op. cit. p.8.

[18] See HDN Guidance for responses...op. cit. page vi.

[19] Grosh M. et al. For Protection...op. cit., Chapter 10.

[20] Africa, Wodon Q., Zaman H. Rising Food Prices in Sub-Saharan Africa: Poverty Impact and Policy Responses. World Bank, PRWP 4738. October 2008.

EAP, Brahmbhatt M., Christiaensen L. Rising Food Prices in East Asia: Challenges and Policy Options. World Bank. May 2008;

ECA, Lindert K., Schwartz A. "Social Protection Responses to the Global Economic Crisis in ECA." Knowledge Brief 52679. March 2009

ECA, The Crisis Hit Home, 2010;

ECA, Rising Food and Energy Prices in Europe and Central Asia, 2011.

LCR, Rising Food Prices: The World's Bank Latin American and Caribbean Region Position Paper. World Bank. 2008.

MNA, Yemtsov R. The Food Crisis: Global perspectives and Impact on MENA—Fiscal and Poverty Impact, MNSED, June 2008.

WB, IFAD, FAO, Improving Food Security in Arab Countries, January 2009.

SAR. Food Price Increases in South Asia: National Responses and Regional Dimensions. June 2010.

How Ready are Latin American and Caribbean Safety Nets for Food Price Increases?, World Bank, Latin American and Caribbean Region. 2011.

[21] Summarized in Grosh M. et al. Assessing...op. cit.

[22] World Bank, Framework Document for Proposed Loans, Credits, and Grants, in the Amount of US$1.2 Billion Equivalent for a Global Food Crisis Response Program. June 26, 2008.

[23] "Appropriate programs include those which offset the income effect of the shock directly through cash or in-kind transfers; and also those which seek to mitigate its consequences on human development outcomes (such as the nutrition, health and education status of children). Such effects might be generated through reduced complementary feeding of children under 24 months of age; through reduced take-up of health services due to income losses; and due to withdrawal of children from school and increased child labor." (GFRP Framework Document, p. 27)

[24] "In developing such responses the Bank will work closely with WFP to establish an appropriate division of labor. In general, it is expected that WFP will lead on the emergency delivery of food to the worst hit countries while the Bank will normally provide assistance through transfer programs, workfare and health and nutrition programs in country." (Ibid.)

[25] Ibid. p. 86.

[26] Ibid.

[27] For this analysis, all GFRP projects with SSN activities are included in the FY09–11 period, although 5 operations were processed in FY08.

[28] These are Burundi, CAR, Guinea-Bissau, Liberia, Madagascar, Senegal, South Sudan, Tanzania, Comoros, Ethiopia, Guinea, Kenya, Sierra Leone, and Togo.

[29] These are CAR, Comoros, Djibouti, Guinea, Guinea-Bissau, Lao PDR, Liberia, Sudan, and Togo.

[30] These were operations in Djibouti, Ethiopia, Guinea, Guinea-Bissau, Kyrgyz Republic, Liberia, Moldova, Nepal, Nicaragua, Sierra Leone (3 operations), Tajikistan, Tanzania, Togo, West Bank and Gaza, and the Republic of Yemen.

[31] Alderman H. and Bundy D. "School feeding programs and development: are we framing the question correctly?." The World Bank Research Observer. Advance Access July 26, 2011.

[32] World Food Progam, Vouchers and Cash Transfers as Food Assistance Instruments: Opportunities and Challenges. Rome, 2008.

[33] For a good summary of the cash versus in-kind transfers see, Grosh M., et al. For Protection...op. cit. Chapter 7.

[34] For a summary of issues and research results on food for education programs in developing countries see Alderman H., Bundy D. "School Feeding Programs ... op. cit..

[35] World Bank, Resilience, Equity, and Opportunity... op. cit. Version March 14, 2012. pp. 45–46.

[36] All scheduled to close either 2012 or 2013.

[37] The latest RSR reports show 16 nutrition activities representing 19% of the RSR portfolio.

[38] For detailed description of nutrition related activities in each country see RSR website and donor progress reports.

[39] The definition of SSN AAA in this section includes SSN work coded 54 (SSN), 91 (food crisis), and AAA with poverty codes under the purview of the SP Sector Board.

[40] Unless explicitly mentioned, this section excludes GFRP SSN operations (33) and commitments ($523 million).

[41] IEG SSN Evaluation. p. 10.

[42] Ibid. pp.13–14.

[43] Specifically in the Africa, SAR, and EAP regions.

[44] IEG SSN Evaluation, p. 18. What distinguishes LICs from fragile states is that the latter have greater governance challenges and additional vulnerabilities such as large numbers of orphans, displaced populations, etc.

[45] WB, World Development Report 2011. Conflict, Security and Development. 2011.

[46] As indicated in the IEG SSN Evaluation, greater ongoing lending to MICs provides additional resources (for SSN loan preparation and supervision) with which to engage in dialogue and further stimulate country demand for SSNs. In fact, a significant predictor of Bank SSN lending is the volume of prior SSN lending.

[47] The RSR was created as part of the Vulnerability Financing Facility, a framework established by the Bank in April 2009 to streamline crisis support to the poor and vulnerable and to provide a broad range of technical and financial assistance to LICs. The facility was designed to address two specific areas of vulnerability to crisis: agriculture (under the GFRP umbrella) and safety nets, employment, and protection of basic social services (under the RSR Program).

[48] See Rapid Response Monthly Report—June 2009.

[49] IEG SSN Evaluation...op. cit. pp. 14–15.

[50] IEG SSN Evaluation...op. cit., pp. 14–15.

[51] IEG RAP 2012 includes only 4 DPOs under SP.

[52] The classification of SSN instruments or type of intervention follows the coding in the IEG SSN Evaluation.

53 See analysis in Sabates-Wheeler R., Devereux S. "Cash Transfers and High Food Prices: Explaining Outcomes on Ethiopia's Productive Safety Net Programme." Future Agricultures Working Paper. London, 2010. When the PSNP was launched in 2005, the cash transfer was set equivalent to the cost of a standard food package. Ethiopia's high inflation rates, especially since 2007 have reduced the real purchasing power of PSNP cash payments relative to the food transfers.

54 There are definitional differences between IEG estimates and the HDN portfolio figures. These numbers are consistent with the social safety net portfolio definition used in this evaluation and the previous IEG SSN evaluation and include social safety net projects with child health activities.

55 See, Ibid.

56 These are projects with thematic code 68 (nutrition and food security) but not code 54 (SSNs) under any SB (they are mostly under HNP and SP).

57 HDN nutrition group estimates include projects managed under all the networks, while IEG definition covers only projects managed by HNP and SP networks.

58 Tanzania, Togo, and Burkina Faso.

59 Bangladesh and Nepal.

60 More details on SSN building blocks in my AAA paper.

61 Such as the national targeting systems in Chile and Colombia, which are used for many of the poverty targeted programs and subsidies, for example, or the unique beneficiary registry in Brazil.

62 The share of regular SSN projects with institutional development components in Africa and EAP declined from 63% and 75% respectively in the pre-crisis period to 46% and 56% respectively in the post-crisis period.

63 HDN. Rapid Social Response Program Progress Report 2012. World Bank, March 2012, p.3.

64 That is all projects overseen by the SP SB including codes 51 (Labor Markets), 54 (SSNs), 55 (Vulnerability Assessment & Monitoring), 56 (Other SP & Risk Management), and 87 (Social Risk Mitigation).

65 It should be noted that the IEG RAP 2012 findings are not comparable to those of the IEG SSN Evaluation of projects 2000–10 as they focus on different portfolios: (a) RAP includes the entire SP portfolio, the SSN Evaluation includes only SSN projects; (b) RAP includes projects based on their "exit" year while the SSN Evaluation included them based on "approval" year; (c) the SSN Evaluation results refer to the entire 2000–10 decade while the RAP 2012 compares FY2006–08 and FY2009–11; (d) the SSN Evaluation does not include AF and SF operations while the RAP includes all completed operations. On RAP 2012 methodology see IEG Results and performance of the World Bank Group 2012. Volume II: Appendixes, August 1, 2012. For IEG SSN Evaluation… op. cit. methodology see Appendix B & IV.

66 It would be useful to get budget data for SP similar to the data used in the agriculture chapter. Major reallocation of resources across Bank sectors or between FACs and Operations are not likely to be large in the short-run due to budgeting mechanisms and processes as well as high fixed costs in FACs units.

67 Rethinking School Feeding, WFP and World Bank, 2009.

# 5

## Lessons and Recommendations

This report on the evaluation of the Bank Group response to the global food crisis has focused on key aspects of the design, implementation, and early outcomes of that response. This chapter distills the lessons learned from and issues raised by the analysis to produce recommendations that will help the Bank Group in responding to future food price crises.

The further evolution of the food crisis is still unclear, food prices are spiking again for the third time in past five years. The prices of internationally traded maize and soybeans reached all-time peaks in July 2012, following an exceptionally hot and dry summer in both the United States and Eastern Europe.[1] Wheat prices have also soared to levels comparable to the 2011 peak but below all-time records. Prices of rice remain stable from abundant supplies. As shown in this report, countries in the Middle East and North Africa and Sub-Saharan Africa regions are most vulnerable to this type of global food price shock. These countries have large food import bills, their food consumption is a large share of average household expenditures, and they have limited fiscal space and weak social safety net systems. These factors, together with widening rural-urban disparities, increase the risk of social and political tensions. These developments have implications for the demand for Bank Group financing, and the Bank Group may need to respond urgently to severe price shocks in the poorest countries in the future. While implementation of the crisis response program of 2008 encountered a number of difficulties highlighted by this evaluation, it also provided insights that have already been internalized. Consequently, the Bank is now in a better position to assist low-income countries in the event of future crises, as noted below, and can further improve its capacity in this regard with the incorporation of additional lessons highlighted in the evaluation.

The implementation of the short-term support program helped build experience for broader institutional crisis response mechanisms. In the past few years, the Bank Group introduced several new instruments to mainstream some of the lessons learned from the GFRP, including the IDA Crisis Response Window and IDA Immediate Response Mechanism. These facilities improved the Bank Group's crisis preparedness. For example, the Bank was very quick to activate the crisis response program in the Horn of Africa using the Crisis Response Window in 2011.

The World Bank Group response program in May 2008 was unique among global financial institutions in speedily articulating a comprehensive, concrete, and fast-disbursing financial support program to provide hard-hit clients with a menu of options for crisis mitigation. Along with Bank Group's longer-term regular agricultural and social protection programs, and knowledge-based policy advice, the GFRP helped solidify the Bank's place as a key player in food security matters. The Bank's constructive participation in the UN High-Level Task Force and contribution to G-7 and G-20 meetings helped the international community to initiate several food security programs. One of these, the Global Agriculture and Food Security Program (GAFSP) arose from the G-20 Summit in September 2009 at which the Bank was asked to prepare a multilateral system to help implement pledges to long-term food security made at the L'Aquila Summit in July 2009. The GAFSP has been operational since April 2010 (see Box 1.1). Another program, the Agricultural Market Information System (AMIS) builds on the recommendations of a joint report[2] and was created by the G-20 Summit in 2011. This program, which works to improve the transparency of international grain physical stocks and markets, also is up and running with major World Bank participation. Twenty-eight countries participate in AMIS.

## General Lessons

Emerging from the evaluation are valuable lessons that can be used to further improve the Bank Group's capacity to address future crisis responses. Five lessons stand out:

- **First, a detailed strategic framework for the Bank Group's crisis response is necessary but not sufficient for the effectiveness of the interventions.** This evaluation found that the GFRP Framework Paper was an important conceptual tool for organizing the

Bank Group's response. However, there was often a disconnect between the intent of policy prescriptions in that paper and what was actually implemented, especially in the short-term fast-tracked programs.

- **Second, enhanced administrative resources—either incremental or redeployed from other purposes—and internal strengthening and collaboration are essential to an effective response that involves an expanded scale of operations.** This lesson is reflected in the evaluation findings for both the Bank and IFC. For the Bank, fast processing had a cost for design quality, implementation, and results in some emergency operations. Moreover, launching such an ambitious crisis response program without an adequate increase in the operational budget and staffing prevented and undermined the quality of existing lending and nonlending operations and had adverse effects on staff work-life balance. IFC's response benefitted from the creation of a variety of trade finance facilities earlier in the decade; however, to some extent the benefits were limited initially by coordination problems across IFC units and between headquarters and regional offices. Subsequent consolidation of three investment departments and significant decentralization mitigated these problems.

- **Third, limited additional resources and pre-crisis IDA allocations can constrain the ability of the Bank to respond to the crisis in IDA-eligible countries.** Beyond the $200 million Food Price Crisis Trust Fund, the Bank Group did not secure additional funding to respond this crisis, and consequently adjustment in assistance to many countries was constrained by IDA allocations that had been determined by criteria unrelated to the crisis, and by limited flexibility within the ongoing country program. For most countries this resulted in modest-size operations that could not have a significant impact on food prices. This experience led to the establishment of Crisis Response Mechanisms, referred to above, that allow IDA countries access to resources beyond their standard IDA allocations.

- **Fourth, the effectiveness of increased lending—as seen in the case of agriculture —depends critically on adequate analytical work and staffing.** The crisis led to greater Bank Group emphasis on agricultural lending. But that emphasis was not supported by the increased staffing, analytic effort, and resources for portfolio management needed to ensure the quality and results of the new and ongoing operations in the sector.

- **Fifth, in countries where social safety net systems are already in place, they can be critical to protecting vulnerable households and individuals during a crisis, but these are rarely in place in low-income countries and fragile states.** As indicated in earlier IEG evaluations (Global Economic Crisis Response and Social Safety Nets), the Bank has done a good job in supporting social protection in middle-income countries, matching growing country demand with innovative approaches and solutions. Although clearly established as a key priority for the new Social Protection and Labor Strategy, what clearly emerges from this evaluation is that the development of feasible approaches in the Bank's tool kit for use by fragile states and the low-income countries is a work in progress. This remains a priority for the Bank's social protection team, with feasible programs included in Country Assistance Strategies, thereby positioning countries to respond to future shocks.

The findings also point out specific lessons for the Bank Group's short-term response.

## Additional Lessons

**Senior management pressure to deliver particular crisis programs carries the risk distorting program composition.** The intensive promotion of the emergency program led to the inclusion of activities not addressing the crisis.

**Pre-existing country-owned agendas and ongoing programs can provide effective platforms for emergency operations.** Building on a pre-existing government-owned agenda and the Bank's strong analytical work, the Philippines GFRP DPO achieved all of its short-term outcomes while catalyzing progress on the longer-term social protection agenda, including establishment of an improved and expanded conditional cash transfer program.

**Context is important in considering the wisdom of tax and tariff reduction in a crisis response. A cautious approach to tariff and tax reductions as part of crisis response is warranted, balancing likely pricing effects with possible implications for fiscal stress.** In many cases, tariffs and taxes on staple foods were low to begin with, and rate reductions did little to help vulnerable groups in alleviating hardships of the food crisis, while aggravating the fiscal situation and threatening other government programs. Some emergency support compensated for budget shortfalls, but typically there was no a priori analysis to advise governments on the utility (and risks) of their tax and tariff policies.

**Good quality results frameworks and monitoring and evaluation arrangements for emergency operations are essential.** The evaluation identified quality risks and concerns in results frameworks of GFRP operations (in both project lending and DPOs), especially the crisis support took the form of additional and supplemental financing arrangements. The latter often bore little substantive relationship to their "parent" operations, but were not uniformly referred to in supervision reports pertaining to the overall project, thus missing opportunities for identifying emerging impacts (and problems) and the need for remedial action. There were also problems in monitoring and evaluation, where, in several cases, monitoring surveys conducted after the closing of operations found evidence of sizeable leakages, as a number of beneficiaries targeted under program and included in the distribution lists had not received food packages or had received incomplete packages at the time when they were interviewed.

**Simple, tried-and-true nutrition and health interventions are essential complements of social safety net programs in a food price crisis and deserve wider use.** The Bank's response to the food crisis has had limited focus on targeting nutrition interventions, with Bank programs in only four low-income countries appear to have emphasized nutrition support to children under age of two and pregnant and breastfeeding women as part of their food crisis response program.

**Effective partnerships at the country level play a vital role to successful implementation of crisis-response programs.** The donor coordination involved in Ethiopia's Productive Safety Nets APL II was unique. In an effort to move to more predictable support and reduce fragmentation in humanitarian support, partners pooled their funds and came together in a unified stream of technical assistance supporting the government-led program. IDA provided additional financing of $25 million to maintain adequate coverage in 2009. But partnerships were also important in countries where the authorities provided less leadership and the risk of donor fragmentation and duplication was greater—in these cases effective communications across donor groups and agencies is even more important for results. In Nepal, WFP has been a key partner in the implementation of Rural Community Infrastructure Public Works program and the food-for-work program. In Tajikistan, Emergency Food Security and Seeds Imports Project focused on the distribution of fertilizer and seed to 28,000 targeted small-scale and poor farmers based on an earlier program funded by the FAO.

## Recommendations

The findings suggest four main recommendations to improve Bank Group effectiveness in responding to food crises.

**First, when the Bank decides to respond to similar crises in the future, ensure that country-driven food crisis response programs are adequately resourced with administrative budgets to facilitate effective preparation and supervision of food crisis mitigation operations.** The GFRP Framework Paper was an important conceptual tool for organizing the Bank Group's response, but implementation encountered problems. Operational resources were not expanded sufficiently for preparation and supervision to match the increased and accelerated volume of operations with adverse consequences for the quality of operations and staff work-life balance, and at the risk that other country priorities would be neglected.

**Second, develop quality assurance procedures for food crisis response programs that mitigate the potential adverse effects of speedy preparation and implementation.** Specifically, the Bank needs to: (a) strengthen ex-ante quality assurance oversight for food crisis response programs prepared under accelerated preparation procedures. Such oversight would ensure, *inter alia*, better alignment between the design of operations and the Bank's food crisis-related policy advice at times of spiking food, fuel and fertilizer prices, particularly with respect to taxes, tariffs, subsidies, and their targeting, considering the country contexts. (b) ensure that food crisis response components, processed as re-structured projects, additional or supplemental finance operations, include appropriate monitoring and evaluation arrangements. (c) require specific reporting on the crisis response components of restructured, additional or supplemental finance projects in implementation status reports, implementation completion reports and other project reports.

The Bank's fast processing of crisis response operations exacted a cost for design quality, implementation, and results in some emergency operations suggesting that additional oversight of the standard quality assurance procedures was needed. In some food crisis response operations, the Bank acquiesced with, or supported, policies and actions that were inconsistent *with its own* food crisis-related policy advice or that were not aligned with the country context. For example, in many countries, tariffs and taxes on staple foods were low to begin with and rate reductions did little to help vulnerable groups while aggravating the fiscal situation and threatening other government programs. In input subsidy operations, the underlying policy rationale was to stimulate a supply response to mitigate the adverse effects of input and food price increases, but the targeting was not consistently conducive to maximize supply response. The presence of other constraints (such as limited supply of quality seeds) was not always considered. Furthermore, the coverage of input subsidy operations was often too small to generate a significant supply response at the national level. Where additional or supplemental finance instruments were used, the monitoring and evaluation arrangements, and the reporting on implementation and results did not consistently cover the food crisis response components of the project, limiting the potential for remedial steps and hindering impact assessment.

**Third, assist countries to better target the people most vulnerable to a food price crisis (especially children under two and pregnant and breastfeeding women) with adequate nutrition interventions in their mitigation programs.** Few Bank programs, in either low- or middle-income countries, emphasized nutritional support to children under age two and pregnant and breastfeeding women (the most vulnerable people) as part of their food crisis response program, even though most countries "vulnerable" to the food crisis have the highest global malnutrition burdens. Thus only a handful of low-income countries (Kyrgyz

Republic, Lao PDR, Liberia, Moldova, Nepal, Sierra Leone, Senegal, and Tajikistan) focused on infant and maternal nutrition in their crisis response. Likewise, only a few middle-income countries emphasized infant and maternal nutrition in their crisis response.

**Fourth, work with client countries and development partners to identify practical mechanisms (including indicators) for monitoring nutritional and welfare outcomes and impacts of food crises and mitigation programs, and work with them to implement those mechanisms and to report the results.** The main welfare outcomes from the crisis—poverty and malnutrition—were not sufficiently tracked to assess the welfare impact of the short-run response. While theory and the Bank's policy guidance provide a framework to assess the impacts of food crisis on the welfare and nutritional status of key population groups, this requires country-specific assessments. Data scarcity is acute for most of low-income countries. Thus, few GFRP countries assessed the impact of the food crisis on the poor. Some social safety net projects under the GFRP described mechanisms for the selection of beneficiaries, mostly using a combination of geographic and then community targeting, a practical approach that can produce serviceable targeting in data-constrained environments. However, the majority of projects did not specify actual and expected program "coverage" to assess the likely contribution of the project to the population in need of assistance. Most project documents state that project activities were targeted to food-insecure areas, but indicators only provide numbers of children to receive food in school or number of hospital patients to be fed.

## Endnotes

[1] Food Price Watch, World Bank, August 2012.

[2] Price Volatility in Food and Agricultural Markets: Policy Recommendations. Policy Report 2011. FAO, IFAD, IMF, UNCTAD, WFP, the World Bank, the WTO, IFPRI and UN HLTF.

# Appendix A
## Differential Effects of the Crisis

Countries that are net food exporters will experience improved terms of trade, while countries that are net food importers will face more challenges meeting domestic demand. However, the number of food importer countries is four times that of food exporter countries, and most countries in Sub-Saharan Africa are net cereal importers (IFPRI 2007). Large net importers of food, such as those in the Middle East and North Africa and in West Africa, face higher import bills, reduced fiscal space, and greater transmission of world prices to local prices for imported rice and wheat (GMR 2012). According to an FAO estimate, Africa's cereal import bill was at about $21.748 billion in 2008 and about $9.8 billion in Sub-Saharan Africa alone in 2008, translating into 30 percent and 35 percent increases over the 2007 level, respectively (IMF 2008, FAO 2008c).

Developing country markets often lack the capacity to absorb domestic shocks, and can be subject to high domestic price volatility. The high and volatile prices at the national and local levels can have dire consequence for poor people, including rural smallholder farmers. The poorest people spend roughly three-quarters of their income on staple foods (Cranfield, Preckel, and Hertel 2007). While the incomes of farm households may be increased by higher commodity prices (Hertel, Ivanic, Preckel, and Cranfield 2004), the benefit to poor farm households may be offset by reduced net sales of these goods (Ivanic and Martin 2008).

As poor households have less means to cope, a food price spike, even short-term, could have strong negative long-term effects on poverty due to the lower caloric intakes and increase in child malnutrition or school dropouts as well as to selling their productive assets including seeds and livestock (Jalan and Ravallion 2002). The impacts of higher food prices on poverty depends upon the reasons for the price change and the structure of the economy, such as the distribution of net food buyers and net food sellers, particularly among households around the poverty line (Hertel and Winters 2006, Ravallion and Lokshin 2005). In low-income countries, poverty increases are considerably more frequent, and larger, than poverty reductions when prices of staple foods are higher (Ivanic and Martin 2008). Poverty and food insecurity often cause degradation of natural resources, and degraded natural resources further worsen poverty and food insecurity, forming a vicious cycle (Pinstrup-Andersen and Herforth 2008).

Finally, rural and urban households and net food sellers and net food buyers experience different effect of a food price spike. In urban areas, higher food prices may hurt most households as they are net food buyers. In rural areas, higher food prices may also hurt as those rural poor who have land can be net food sellers or net food buyers. While larger farmers and those small holders who are net sellers of food will benefit from food crops increases, a large share—in many countries over 50 percent—of rural households are net food buyers, and hence are likely to be adversely affected by the food price spike (Christiansen and Demery 2007). In rural Bangladesh, for example, 80 percent of the poor are smallholders and 62 percent of the poor smallholders are net buyers of food. In other words, half of the country's poor are net food buyers, and hence suffered from food price increase (de Janvry and Sadoulet 2011), not only because food consumption stands for a larger share of total consumption of poor households, but also because the overall inflation depletes their already meager assets more severely (Ravallion and Datt 2002, Ivanic and Martin 2008).

An increase in international food prices would generally be expected to generate incentives to farmers their production, with some delay due to the seasonality of agriculture production. If the increase in global prices coincides with an increase in the price of key agricultural inputs such as fertilizer and fuel (as was the case in the 2007–08 food crisis), the improvement in incentives due to higher output prices will be diminished, and the supply response would be subdued. The extent of transmital of global prices to domestic markets differs across countries, depending on factors such as transport costs and the extent of substitutability between domestically produced staples and imported staples, but in many countries, transmital is only partial (Minot, 2011). Furthermore, in most countries facing a food crisis, governments undertake remedial short-term policies designed to lower domestic food prices to consumers, often using measures that depress the prices that would otherwise be faced by producers (e.g., bulk release of strategic food reserves, tariff reduction, price controls, export bans or other limitations on exports).

TABLE A.1  Factors in the Global Food Crisis

| Time Horizon | Demand-Side Factors | Supply-Side Factors |
|---|---|---|
| **Long Run** | Increasing population.<br><br>Rising incomes in developing countries leading to increased consumer purchasing power, increased demand for meat and dairy products, and increased indirect demand for grains. | Limited availability of agricultural land and water for irrigation; insufficient investments in rural institutions and infrastructure, agricultural research, extension, and water and soil management; poor policies in some developing countries; Organisation for Economic Cooperation and Development subsidies; climate change; inadequate systems to ensure food safety. |
| **Medium Run** | Biofuel demand. | Rising energy prices and resulting increases in prices of fertilizers, pesticides, and transportation; subsidies for biofuel production. |
| **Short Run, Cyclical** | Financial speculation that may exacerbate the price effects of food shortages. | Adverse weather in major exporting countries, crop diseases, exchange rate volatility, price controls and changes in export and import policies, speculative hoarding, untargeted subsidies. |
| **Recent** | Financial crisis and resulting credit tightening and increased borrowing costs for food exports and imports (OECD 2009). | Food security concerns prompting major buyers in the world market (for example, countries in the Middle East and North Africa) to lease land for agricultural production in Sub-Saharan Africa.<br><br>Diversion of land from wheat and other crops to production of biofuel feedstock; increase in farmland process (Von Braun and Meinzen-Dick 2009); low global grain stocks; tighter credit availability for crop production because of the financial crisis (OECD 2009). |

SOURCE: Growth and Productivity in Agriculture and Agribusiness, IEG 2011a. Also see World Bank 2012.

# References

Christiaensen, Luc, and Lionel Demery. 2007. *"Down to Earth: Agriculture and Poverty Reduction in Africa."* Washington, DC: World Bank.

Cranfield, J. A. L., Paul V. Preckel, and Thomas W. Hertel. 2007. *"Poverty analysis using an international cross-country demand system."* World Bank Policy Research Working Paper Series 4285.

Datt, Gaurav, and Martin Ravallion. 2002. *"Is India's Economic Growth Leaving the Poor Behind?"* Journal of Economic Perspectives, American Economic Association, vol. 16(3), 89–108.

De Janvry, Alain, and Elisabeth Sadoulet. 2008. *"The Global Food Crisis: Identification of the Vulnerable and Policy Responses."* UC Berkeley, Giannini Foundation of Agricultural Economics: Agriculture and Resource Economics Update 12 (2), 18–21.

Hertel, Thomas W., Maros Ivanic, Paul Preckel, and John Cranfield. 2004. *"The Earnings Effects of Multilateral Trade Liberalization: Implications for Poverty in Developing Countries."* World Bank Economic Review 18(2): 205–236.

Hertel, Thomas W., and L. Alan Winters (editors). 2006. *"Poverty and the WTO: Impacts of the Doha Development Agenda."* New York: Palgrave MacMillan.

Jalan, Jyotsna, and Martin Ravallion. 2002. *"Household Income Dynamics in Rural China."* Discussion Paper No. 2002–10, United Nations University, World Institute for Development Economics Research.

Minot, Nicholas. 2011. *"Transmission of world food price changes to markets in sub-Saharan Africa."* IFPRI discussion paper 1059, January 2011.

Pinstrup-Andersen, Per, and Anna Herforth. 2008. *"Food Security: Achieving the Potential."* Environment 50 (5): 48–61.

Ravallion, Martin, and Lokshin, Michael. 2005. *"Lasting Local Impacts of an Economy Wide Crisis."* World Bank Policy Research Working Paper Series 3503.

Von Braun, Joachim. 2007. *"The World Food Situation: New Driving Forces and Required Actions."* IFPRI's Bi-Annual Overview of the World Food Situation.

Von Braun, Joachim, and Ruth Meinzen-Dick. 2009. *"'Land Grabbing' by Foreign Investors in Developing Countries: Risks and Opportunities."* IFPRI Policy Brief 13. April 2009.

World Bank. 2007. *"Global Development Finance: The Globalization of Corporate Finance in Developing Countries."* Washington, DC.

# Appendix B
# Timeline of the World Bank Group Response and the Response of Other Donors

This appendix describes the chronology of events surrounding the international and World Bank Group response to the global food price crisis and summarizes the response of other donors.

## Timeline of the World Bank Group Response

World Bank research provided early signals that a food price spike was coming. In May 2007, *Global Development Finance* warned that the "reorientation of agricultural output toward biofuels, together with a change in stocking policy in China, has reduced global grain stocks." "Supply conditions are so tight," it said, "that a major supply shock could result in the price of these grains rising much more rapidly, with wheat and maize prices possibly rising more than 40 percent" (World Bank 2007: 32). If such price increases were to occur, the report said, they would have a significant impact on the gross domestic product of many developing countries and a particularly devastating effect on the welfare of the poor, especially the urban poor.

Six months later, in the first week of November 2007, the Food and Agriculture Organization's (FAO) biannual *Food Outlook Report*[1] documented a sharp increase in global food prices (37 percent increase in the global food price index compared to 2006, and even larger increases in key grain prices). The report warned that "developing countries as a whole could face a year of increase of 25 percent in aggregate food import bills. Among them, the most economically vulnerable countries are set to bear the highest burden in the cost of importing food."[2]

On December 11, 2007, a Brief from the FAO's Global Information and Early Warning System on Food and Agriculture[3] reported on food riots in Guinea, Mauritania, Mexico, Morocco, Senegal, Uzbekistan, and the Republic of Yemen, and described the consequences of the escalating food prices for the poor. It listed nearly 20 countries that were already (or likely to be soon) hard hit by high cereal prices. A week later, on December 17, 2008, FAO Director General Jaques Diouf raised the level of alarm in a press conference where he called on the international community to take new steps "to prevent the negative impacts of rising food prices from further escalating and to quickly boost crop production in the most affected countries."[4]

In a major speech on April 2, 2008, World Bank President Zoellick called for a "New Deal for Global Food Policy" to focus not only on hunger and malnutrition as well as access to food and its supply but also on the interconnections with energy, crop yields, climate change, investment, and the marginalization of women.[5] The World Bank Group at the time estimated that 33 countries faced potential social unrest because of the acute hike in food and energy prices. This speech was followed by the report "Rising Food Prices: Policy Options and World Bank Response."[6] That report reviewed policy measures various countries had taken to mitigate the impact of high food prices, and described Bank Group assistance to implement short- and medium-term country-level activities to respond to the crisis, highlighting the provision of policy advice.[7]

The global food crisis was a major theme on the agenda of the Spring Meetings of the World Bank and IMF in mid-April 2008. At those meetings, President Zoellick reiterated his vision of a New Deal for Global Food Policy and won support for the concept at the Development Committee meetings, including endorsement of the strategy of combining immediate financial assistance to hard-hit countries with long-term lending to boost agricultural productivity. Specifically, the president "proposed to expand current programs and offer emergency financing on a quick and flexible basis."[8]

The World Bank Group responded to the global food crisis with a variety of short-term and medium-term interventions. The framework document for the Global Food Crisis Response Program (GFRP), the Bank Group's principal short-term response, cited the Bank Group's multisectoral expertise and its presence in many of the most vulnerable countries as its comparative advantage in providing solutions. Under the framework, the Bank Group would rapidly provide funds to affected countries and access to innovative financial instruments to mitigate a portion of the food price risk. The framework document noted that the Bank Group could undertake policy analysis that draws on country, regional, and global experience, and that it had the capacity to design and deliver targeted social protection programs to mitigate the negative impact of higher food prices on the poor and vulnerable. The Bank Group could also support policy, programmatic, and investment operations to enhance a food production supply response in the short and medium term. Finally, it was expected that through the International Finance Corporation (IFC), the Bank Group could support private sector activities and investments that would alleviate the effects of the crisis.[9]

# The Food Crisis Response of Key Partners

ROME-BASED AGENCIES

The United Nations' three Rome-based food and agriculture organizations were central to the formulation of the global response—and to its implementation at the country level. The Food and Agriculture Organization (FAO) sounded the alarm in mid-2007 about rising grain prices on world markets, and launched its Initiative on Soaring Food Prices in December 2007 to "help smallholder farmers grow more food and earn more money," based on strategies and actions developed by interagency assessments in 58 countries. As the food aid arm of the United Nations system, the World Food Programme (WFP) also provided an early warning about the breadth of the crisis and attracted media attention with the Executive Director's communications calling the crisis a "silent tsunami" and putting a price tag ($755 million) on the impact the crisis would have on the acquisition costs for WFP's upcoming emergency food assistance program. Finally, in April 2008, the International Fund for Agricultural Development (IFAD) announced that it would make available up to $200 million to enable poor farmers to access essential inputs such as seeds and fertilizer, to allow them to prepare for the forthcoming cropping season as well as to establish a basis for sustained increases in production in subsequent seasons.

## INTERNATIONAL MONETARY FUND

The International Monetary Fund (IMF) participated in the High-Level Task Force, but played a limited role because of the nature of the crisis. Three features of its approach are relevant. First, the IMF did not set out a special food security initiative, though it explicitly acknowledged the difficulties that rising food prices were causing for governments and households. In most cases—in the context of programs and Article IV consultations—the IMF lumped the food and fuel "crises" together. It was rare that IMF spoke about the "food crisis" in isolation, either in institutional pronouncements or in country program documents. In terms of financial support to countries dealing with food and fuel price shocks it relied on the Exogenous Shocks Facility, and later on the Stand-by Credit Facility. Second, and in keeping with the "policy advice" entries in the above tables, IMF was pretty forgiving about country policy responses while clearly articulating its preference for narrowly targeted social protection schemes over broad-based subsidies. Finally, it explicitly deferred to the Bank on agricultural policies.

## REGIONAL DEVELOPMENT BANKS

The major regional development banks, though not participants in the High-Level Task Force, did have programs at the regional level. The recent Global Monitoring Report summarized the approaches of the multilateral development banks (MDBs). In brief, it showed that the banks responded to the food crisis through different lending and nonlending mechanisms: "these responses include emergency financial support to the most vulnerable countries, medium-term assistance to strengthen social safety nets and agribusiness, and long-term programs to enhance infrastructure, rural development, and productivity along the food value chain...." The Report also set out the specific programs that each of the banks were supporting through their responses. The details of these programs vary, but they share a common focus on the structural and long-term problems associated with food insecurity, which in turn translated into support for the building and maintenance of rural infrastructure. In addition, the African Development Bank, for example, provided budget support that also supported the strengthening of safety nets. The Inter-American Development Bank also focused, initially, on both short-term safety net support and medium term-production support through its Food Security Fund and other mechanisms; but it was subsequently refocused on agricultural production, productivity, and trade as vehicles for enhancing food security. The Asian Development Bank focused on transportation and communications, as well as water-based natural resource management. The European Bank for Reconstruction and Development focused its efforts on assisting the private sector, largely through helping farmers improve their risk management and access to financing and by supporting infrastructure and trade logistics given their important role in smoothing price fluctuations.

## Endnotes

[1] Food and Agriculture Organization (FAO). Food Outlook Report 2008. Accessed at (http://www.fao.org/docrep/010/ah876e/ah876e00.htm.

[2] In a press conference to introduce the report, the chief of the FAO's Commodity Markets and Policy Analysis Service referred to the situation as a food crisis and highlighted the risk to long-term nutrition as well as the increased likelihood of spreading of social unrest in developing countries, following well-publicized food riots in Mexico and the Republic of Yemen. See coverage of FAO's press conference in the Financial Times article of November 7, 2007, entitled "Rising Food Prices to hit consumption," by Javier Blas, accessed at http://www.ft.com/cms/s/0/6be9fe80-8ca2-11dc-b887-0000779fd2ac.html#axzz1vkWzTQkh.

[3] "High Cereal Prices Are Hurting Vulnerable Populations in Developing Countries," Food and Agriculture Organization (2008), accessed at http://www.fao.org/giews/english/shortnews/highprice071211.htm.

[4] FAO press release, December 17, 2008, accessed at http://www.fao.org/newsroom/en/news/2007/1000733/index.html. In the same press conference, the FAO announced the launch of its Initiative on Soaring Food Prices. Through this initiative the FAO would provide advice and assistance to governments (through specific country programs) on responding to the crisis. Furthermore, it was envisioned that with sufficient donor funds, the FAO would support programs to promote fast supply response among smallholders by providing access to inputs. The FAO allocated $17 million from its existing budget to start these activities. In the coming months, the FAO raised funds from several bilateral and international donors to enable an expansion of its initiative, and the detailed Programme Document for the initiative was only finalized in May 2008. Food and Agriculture Organization (FAO). Initiative on Soaring Food Prices, 2008. Accessed at http://www.fao.org/fileadmin/user_upload/ISFP/ISFP_Programme_Document.pdf.

[5] Financial Times, January 24, 2008, "Zoellick calls for fight against hunger to be global priority," accessed at http://www.ft.com/intl/cms/s/0/b234bf14-ca1f-11dc-b5dc-000077b07658.html#axzz1wC4JMX00.

[6] World Bank Press Release, April 2, 2008, "World Bank President Calls for Plan to Fight Hunger in Pre-Spring Meetings Address." Accessed at http://web.worldbank.org/WBSITE/EXTERNAL/NEWS/0,,contentMDK:21711537~menuPK:34457~pagePK:34370~piPK:34424~theSitePK:4607,00.html.

[7] In introducing the report, President Zoellick emphasized that "In some countries, hard-won gains in overcoming poverty may now be reversed. As an international community we must rally not only to offer immediate support, but to help countries identify actions and policies to reduce the impact on the world's most vulnerable." World Bank, *Rising Food Prices: Policy Options and World Bank Response*, April 2008. Accessed at http://siteresources.worldbank.org/NEWS/Resources/risingfoodprices_backgroundnote_apr08.pdf.

[8] Bank Information Center, May 9, 2009. "Update: "Amid food riots and shaken governments IFIs scramble to develop a coherent response." Accessed at http://www.bicusa.org/en/Article.3763.aspx.

[9] Framework Document for Proposed Loans, Credits and Grants for a Global Food Crisis Response Program (World Bank May 2008).

FIGURE B.1  Timeline of the Crisis and the Response (Calendar Year)

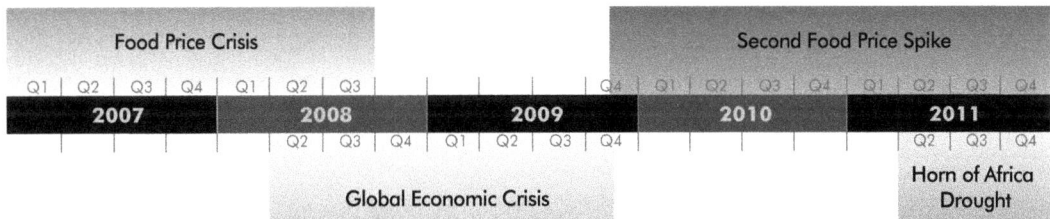

**2007/ Q1**
International grain prices start to rise sharply

**2007/ Q4**
Initiative on Soaring Food Prices (ISFP) announced at FAO

**2007/ Q4**
International grain prices spike

**2008/ Q2**
World Bank launched the Global Food Crisis Response Program (GFRP) up to $1.2 billion

**2009/ Q2**
GFRP increased to up to $2 billion

**2010/ Q2**
Global Agriculture and Food Security Program (GAFSP) was launched to support food security ($20 billion over three years)

**2010/ Q2**
International grain prices spike again

**2011/ Q2**
World Bank extended the GFRP until June 2012. World Bank $500 million support program for the Horn of Africa

**2011/ Q3**
World Bank increased to $1.88 billion its support program to the Horn of Africa

# Appendix C
# Analytical Framework and Selection Criteria for the Case Studies

## Analytical Framework for the Results Chain

The analytical framework for the evaluation, depicted in Figure 1.2, sketches the program theory behind the Bank Group's response to the global food price crisis at the country level, a "results chain" linking inputs and outputs to intermediate and long-term outcomes and impacts.

**Inputs:** Inputs are the special crisis response initiatives and programs launched to mitigate the short-term adverse impacts of the spike in food prices (such as the GFRP and GFI) and the standard Bank Group instruments in agriculture and social safety net sectors that include IDA/IBRD loans, credits, trust funds, World Bank analytic and advisory activities (AAA), and IFC advisory services and investment activities, which were oriented toward improving vulnerable countries' capacity to withstand and manage future crises.

**Outputs:** Outputs are divided into two categories: (a) policies and programs to tackle price volatility and its short-term social and economic impacts, and (b) policies and programs to enhance resilience to future crises in the longer term. The shorter-term measures include policies to stabilize and reduce domestic food prices (such as tariff and tax reductions and targeted price subsidies), introduction and expansion of social safety nets (such as cash and food transfers and school feeding programs), support to the farm sector to induce short-term and medium-term supply response (such as input subsidies and seed distribution), technical and policy advice, and coordination with other donor assistance programs. Speed of preparation and implementation were considered as process indicators for the timeliness of response. Longer-term measures include expansion of agricultural production capacity, enhancing productivity growth, developing risk management tools, and further building up the organizational infrastructure and operational capacity of safety net programs.

**Intermediate Outcomes:** Short-term and medium-term crisis response measures translate into intermediate outcomes, such as mitigation of domestic price spikes, channeling significant assistance to vulnerable groups, and increasing domestic supply of staple food products through improving access to inputs. The key indicators for assessing intermediate outcomes are domestic food price index (relative to pre-intervention and international price trends), food crop output volume, and the number of social safety net program beneficiaries as a percentage of targeted vulnerable population. The evaluation also assessed the capacity to identify poor and vulnerable households and to deliver benefits to vulnerable groups as indicative of success.

**Indicative Longer-term Outcomes:** Longer-term resilience enhancement is reflected in expanded food production capacity, increased strategic food reserves, and improved capacity of social safety nets. The key indicators for assessing longer-term resilience enhancement are food crop yields, volume of food crop production, volume of strategic grain reserves, number of social safety net beneficiaries as percent of vulnerable population. In terms of improved capacity of social safety nets, the extent to which institutions in place have the ability to scale up their operations in response to future crises, and availability of contingency funds and donor coordination arrangements are important factors. These resilience indicators correspond to the "availability" and, in part, to the "access" aspects of the concept of food security, the sense that increased agricultural productive capacity improves the availability of food, while improved safety net infrastructure enables governments to expand vulnerable households' access to food in times of crisis through cash and in-kind transfers.

## Evaluation Building Blocks

**Review of the GFRP Lending Portfolio:** The portfolio consists of 55 operations in 35 countries. Business Warehouse has been complemented with the addition of variables based on in-depth review of program and project documents to determine the type of support provided in response to the food crisis and the results achieved. (Support might include, for example, short-term immediate support such as school feeding programs or providing farmers with agricultural inputs or longer-term capacity building such as developing surveys to measure household welfare or measures to increase farmers' productivity).

**Review of IFC's Food Crisis Response:** The evaluation investigates the objectives of IFC's financial investments in GFI, based on a modified version of its standard evaluation methodology, adapted to capture dimensions important to the food crisis, such as relevance, speed of response, systemic impacts, outputs and preliminary outcomes, and IFC's role and contribution. As part of the short-term food crisis response, IFC offered advisory services under the GFI program and mobilized $300 million for advisory services activities. The evaluation reviews the scope and effectiveness of the food crisis advisory services using established IEG criteria, modified to reflect the food crisis response backdrop, focusing at the present stage on outputs and preliminary outcomes, as opposed to final outcomes or impact.

**Review of the Agriculture and Social Safety Nets Lending Portfolio:** The review of the World Bank's agriculture and social safety net lending portfolio is one of the main instruments to provide comprehensive insights on the Bank Group's medium and longer-term food crisis response as it goes beyond the GFRP operations. The portfolio review examines trends in Bank engagement over time and throughout the regions. The portfolio review looks at the extent to which the focus of Bank Group lending in agriculture and social safety nets has shifted toward building resilience in the aftermath of the crisis.[1] It builds on the databases and analyses prepared by two previous IEG evaluations: the IEG evaluation of Social Safety Nets and IEG evaluation of Growth and Productivity in Agriculture and Agribusiness. Both portfolios were updated by adding FY2011 operations. A similar portfolio analysis was carried out for IFC's medium- and longer-term crisis response.

**Review of Agriculture and Social Safety Nets Analytical and Advisory Services:** The review included economic and sector work and technical assistance classified in the Bank's internal database (Business Warehouse) as food crisis response as well as relevant agriculture and safety net reports. The latter includes activities completed by the Bank between FY2007 and FY2011.

**Country Case Studies (Field and Desk-Based Country Cases):**[2] Country case studies were used to assess the extent to which the Bank Group and GFRP/GFI response was relevant and effective, the coordinated efforts of the donor community effectively addressed the short- and longer-term objectives, vulnerable countries were able to set up systems to address future food price shocks, and the constraints were removed to a more effective response.

## Selection Criteria for the Country Case Studies

The selection of the 20 countries to study in-depth took into account those most severely affected by the crisis and those that received most GFRP and GFI funding. Field work was carried out for 9 country case studies. The case studies reviewed the broader Bank Group safety net and agriculture operations and analytic and advisory activities. Sample selection was purposive, based on two criteria:

- **Operation size:** The GFRP/GFI involved both large (between $10 million and $275 million) and small operations (less than $7 million). The top five GFRP borrowers (Ethiopia, Tanzania, Philippines, Bangladesh, Nepal), accounting for 60 percent of the funds committed, were all selected for in-depth review.

• **Regional and sector balance:** Another 15 countries were selected to achieve a combination of regional and sector balance. Selection was proportional to the regional distribution of the GFRP countries (Table C.1). Ten countries were selected from Africa, covering the 10 largest operations in the region. At least two countries were selected from each region other than Africa.

## Endnotes

[1] The set of agricultural-oriented operations covered in evaluation includes all operations under the Agricultural and Rural Sector Board and other Sector Boards that have agricultural components or rural infrastructure (agricultural extension and research, crops, irrigation and drainage, animal production, general agriculture and fishery, public administration-agriculture, agro-industry, agro-industry marketing and trade, rural land administration, rural infrastructure such as rural roads and rural energy). Similarly, SSN will include all projects approved by the Social Protection Sector Board with theme code 54 (social safety nets) as well as all projects approved by other Sector Boards assigned theme code 56 (other social protection and risk management).

[2] Two Project Performance Assessment Reports (PPAR) have been completed on GFRP operations in Ethiopia and another two are in advance stages for GFRP operations in Burundi and Djibouti.

TABLE C.1  Regional Breakdown of the Case Studies

|  | Africa | East Asia and Pacific | Eastern Europe and Central Asia | Latin America and Caribbean | Middle East and North Africa | South Asia |
|---|---|---|---|---|---|---|
| Percentage of Case Studies | 50 | 10 | 10 | 10 | 10 | 10 |
| Percentage of GFRP Countries | 57.5 | 8.5 | 8.5 | 8.5 | 8.5 | 8.5 |

**BOX D.1** Policy Options in the GFRP Framework Document

1. FOOD PRICE POLICY AND MARKET STABILIZATION

A. **Food Price Policy: Crisis Options, Transition, and Longer-Term Approaches**

A.1 Rapid assessment and analytical support to provide a rapid diagnostic of current market conditions, impacts of the crisis on different social groups, and policy options.

A.2 Design of national food policies that provide for a transition from short-term emergency measures to policies consistent with a long-run growth and poverty reduction strategy.

A.3 Information, consultation, and participatory advisory services.

B. **Support for Food Market Stabilization**

B.1 Tax and trade policies, involving tax and tariff reduction that entail government revenue loss and a need for budget support.

B.2 Price subsidies on food, administered in various formats, and requiring budget support to offset the associated financial costs.

B.3 Grain stock management, entailing various forms of technical assistance to governments contemplating expansion or acquisition of grain stocks into their strategic reserves.

B.4 Price risk management, entailing various forms of technical assistance to governments considering utilization of market-based instruments for price hedging, and partial financing of premiums.

B.5 Early warning and weather risk management for food crop production, entailing financing of technical assistance and the set-up of appropriate systems.

B.6 Promotion of bilateral or regional trade, entailing the financing of related technical assistance and infrastructure investments.

*continued*

## 2. ENHANCING DOMESTIC FOOD PRODUCTION AND MARKETING RESPONSE

### A. Strengthening Agricultural Production Systems

A.1 Improving smallholder access to seed and fertilizer through technical assistance, policy reform affecting input production, supply, and credit arrangements, and financing "smart" subsidies to eligible farmers.

A.2 Livestock management for vulnerable households, through the financing of services, Infrastructure, subsidized inputs, and technical assistance for medium-term planning for sustainable resource management.

A.3 Rehabilitation of small-scale irrigation entailing financing of infrastructure, training, and studies to underpin medium term irrigation investments.

A.4 Strengthening farmer access to critical information through the financing of extension and related information diffusion activities.

### B. Reducing Post-Harvest and Marketing Losses

B.1 Reduction of post-harvest losses through the financing of low cost on farm and community storage technologies and facilities; rehabilitation of rural transport; training of regulators, processors, and wholesalers; and investments to upgrade their facilities.

### C. Strengthening Access to Finance and Risk Management Tools

C.1 Improvement and expansion of credit availability to agricultural *producers*, food processors, and traders, through the financing of credit lines to, and capacity building in, formal and community-level financial.

## 3. SOCIAL PROTECTION ACTIONS

### A. Rapid-Response Diagnostics

### B. Financing of Short-Term Support to the Most Vulnerable Populations

B.1 Transfer program (cash transfer, food stamp, food rations)

B.2 School feeding

B.3 Public works

B.4 Nutrition and health programs

### C. Strengthening Social Protection Programs

SOURCE: GFRP Framework Document.

TABLE D.1  Use of Suggested Design Options in GFRP Operations

| Component I | Component II | Component III |
|---|---|---|
| **Price Policy and Market Stabilization** | **Social Protection to Ensure Food Access and Minimize the Nutritional Impact of the Crisis on the Poor and Vulnerable** | **Enhancing Domestic Food Production and Marketing Response** |
| **Tax and Trade Policies**<br><br>Burundi, Djibouti, Guinea, Madagascar, Sierra Leone | **Unconditional Cash or Near Cash Transfers**<br><br>Ethiopia, Kyrgyzstan, Senegal, Tanzania, West Bank and Gaza, the Republic of Yemen | **Improving Smallholder Access to Seed and Fertilizer**<br><br>Bangladesh, Benin, Cambodia, Central African Republic, Ethiopia, Guinea, Honduras, Kenya, Kyrgyzstan, Lao PDR, Liberia, Madagascar, Mali, Nepal, Nicaragua, Niger, Rwanda, Somalia, South Sudan, Tanzania, Tajikistan, Togo |
| **Price Risk Management**<br><br>Haiti, Philippines | **In-kind Transfer and Food-Based Programs**<br><br>Bangladesh, Burundi, Cambodia, Central African Republic, Djibouti, Guinea-Bissau, Haiti, Honduras, Kenya, Kyrgyzstan, Liberia, Madagascar, Moldova, Nepal, Nicaragua, Philippines, Senegal, Sierra Leone, South Sudan, Tajikistan, Togo | |
| | **General Price Subsidies**<br><br>Bangladesh | **Rehabilitation of Small-Scale Irrigation Systems**<br><br>Afghanistan, Mozambique, Nepal, Senegal, Tanzania |
| | **Public Works—Employment**<br><br>Bangladesh, Cambodia, Comoros, Ethiopia, Guinea, Guinea-Bissau, Liberia, Madagascar, Sierra Leone, the Republic of Yemen | |
| | **Conditional Cash Transfers**<br><br>Kenya, Lao PDR, Philippines | |

TABLE D.2 List of 55 Global Food Response Program Projects

| Approval FY | Project | | Economy | Region | Sector Board |
| | ID | Name | | | |
|---|---|---|---|---|---|
| 2009 | P107313 | Fifth Poverty Reduction Support Credit (PRSC-5) | Mozambique | Africa | Economic Policy |
| 2009 | P111545 | Kenya Cash Transfer for Orphans and Vulnerable Children | Kenya | Africa | Social Protection |
| 2008 | P112017 | Food Crisis Response Development Policy Grant | Djibouti | Middle East and North Africa | Economic Policy |
| 2009 | P112023 | Food Prices Crisis Supplemental to Honduras First Programmatic Financial Sector Development Policy Credit | Honduras | Latin America and the Caribbean | Financial and Private Sector Development |
| 2008 | P112083 | Agricultural Infrastructure and Development Project—Additional Financing | Liberia | Africa | Transport |
| 2008 | P112084 | Additional Financing to CEP—Public Works Program | Liberia | Africa | Social Protection |
| 2008 | P112107 | Liberia Emergency Food Support for Vulnerable Women and Children | Liberia | Africa | Agriculture and Rural Development |

| Project Development Objectives | Lending Instrument | Financing Instrument | Loan Amount (US $ millions) |
|---|---|---|---|
| The PRSC-5 also supports the implementation of the government policy response to the higher global food and fuel prices. | DPL | New | 20 |
| The PDO is to increase social safety net access for extremely poor orphans and vulnerable children (OVC) households, through an effective and efficient expansion of the cash transfer OVC program. | IL | New | 50 |
| The objective of the grant is to mitigate the impact of the food crisis on the poor while maintaining fiscal stability. Achieving these goals will reduce hunger, avoid an increase in poverty, prevent social unrest and ensure the fiscal stability needed to foster social and economic development. | DPL | New | 5 |
| The proposed operation would support the government's commitment to maintain macroeconomic stability and persevere in its financial sector development policy credit's development objectives and allow the government to respond to the food price crisis. As such, the supplement will be processed under GFRP procedures. | DPL | SF | 10 |
| The project will support the government's efforts in re-establishing basic infrastructure and reviving the agriculture activities. | IL | AF | 3 |
| As part of the government response to the social and economic emergency of Liberia, the project will improve poor rural communities' access to basic infrastructure and provide economic opportunities for vulnerable households in urban and rural areas. | IL | AF | 3 |
| Maintain access to food among vulnerable households. | IL | New | 4 |

| Approval FY | Project | | Economy | Region | Sector Board |
|---|---|---|---|---|---|
| | ID | Name | | | |
| 2008 | P112133 | Supplemental Second Economic Governance Reform Operation (EGRO-II) Development Policy Grant | Haiti | Latin America and the Caribbean | Economic Policy |
| 2008 | P112136 | Community and Basic Health—Additional Financing | Tajikistan | Europe and Central Asia | Health, Nutrition, and Population |
| 2008 | P112142 | Health and Social Protection Project | Kyrgyz Republic | Europe and Central Asia | Health, Nutrition, and Population |
| 2008 | P112157 | Emergency Food Security and Seed Imports Project | Tajikistan | Europe and Central Asia | Agriculture and Rural Development |
| 2008 | P112186 | Agricultural Investments and Services Project—Additional Financing | Kyrgyz Republic | Europe and Central Asia | Agriculture and Rural Development |

| Project Development Objectives | Lending Instrument | Financing Instrument | Loan Amount (US $ millions) |
|---|---|---|---|
| The proposed supplemental financing would support the government's program aimed at: (a) maintaining gains in macroeconomic stability; and (b) sustaining and continuing to make progress in implementing reforms supported under EGRO II, while helping fill the unanticipated financing gap as a result of the food price crisis. | DPL | SF | 10 |
| The proposed additional grant would help finance the costs associated with an additional intervention that will provide nutritional supplements and nutrition education to pregnant and lactating women, infants and small children. This is a scaled-up activity of component c to enhance the impact of the well-performing Community and Basic Health Project. | IL | AF | 4 |
| The revised project development objectives would be to improve health status in the Kyrgyz Republic: (a) by improving access, financial protection, efficiency, equity, and fiduciary performance in the Kyrgyz health sector; (b) to ensure sufficient and reliable financing for the health sector; (c) to strengthen the targeting of social benefits by developing effective administration and information management systems to improve access to social services in general; (d) to protect and improve health and nutritional status of particularly vulnerable populations in the Kyrgyz Republic in the face of food price shocks, by providing nutritional supplements and nutrition education to pregnant/lactating women and infants/young children; and (e) to help poor Kyrgyz families manage and mitigate the impact of food price shocks and protect consumption (general and food consumption) by scaling up and strengthening targeted cash transfers. | IL | AF | 6 |
| The objective of the project is to increase domestic food production and reduce the loss of livestock to help at least 28,000 poorest households in a timely manner to reduce the negative impact of high and volatile food prices. | IL | New | 5 |
| To improve institutional and infrastructure environment for farmers and herders, with a strong emphasis of the livestock sector. More specifically the project will increase farmer's productivity, particularly of livestock farmers in the project areas and reduce animal diseases that have a public health impact. | IL | AF | 4 |

| Approval FY | Project | | Economy | Region | Sector Board |
|---|---|---|---|---|---|
| | ID | Name | | | |
| 2008 | P112345 | Third Social Fund for Development Project—Additional Financing | Republic of Yemen | Middle East and North Africa | Social Protection |
| 2009 | P112761 | Bangladesh Food Crisis Development Support Credit | Bangladesh | South Asia | Economic Policy |
| 2009 | P112908 | Health Services and Social Assistance—Additional Financing | Moldova | Europe and Central Asia | Health, Nutrition, and Population |
| 2009 | P113002 | Social Safety Nets Project | Nepal | South Asia | Agriculture and Rural Development |
| 2009 | P113117 | Food Price Crisis Response Program—Additional Financing to Social Safety Net Reform Project | West Bank and Gaza | Middle East and North Africa | Social Protection |
| 2009 | P113134 | Emergency Food Security and Reconstruction Project | Madagascar | Africa | Social Protection |
| 2009 | P113141 | Additional Financing to NSAP—Food Crisis Response | Sierra Leone | Africa | Social Protection |

| Project Development Objectives | Lending Instrument | Financing Instrument | Loan Amount (US $ millions) |
|---|---|---|---|
| The project development objective (PDO) will remain the same, namely, to improve the range of services and economic opportunities available to the poorer segments of the population through the carrying out of community development, microfinance, and capacity building programs. | IL | AF | 10 |
| The Food Crisis Development Support Credit (FCDSC) was designed to mitigate the impact of high food prices and enhance food security through an expansion of the food-related safety net programs while maintaining sustainable public finances. | DPL | New | 130 |
| The original project development objective will remain the same, that is, to increase access to quality and efficient health services with the aim of decreasing premature mortality and disability for the local population and improve the targeting of social transfers and services to the poor in line with the Medium-Term Expenditure Framework (MTEF) for 2007–09. | IL | AF | 7 |
| To address the short a medium term implications of the global food crisis in the country by providing access to food and strengthening agriculture production, particularly for food insecure districts and small holders. | IL | New | 21.7 |
| The objective is to capitalize on the Bank's Social Safety Net Reform Project (SSNRP) cash benefit scheme to make a one time payment of $100 to around 50,000 households that have been negatively affected by the food crisis. | IL | AF | 5 |
| The project development objectives are to: (a) increase access to short-term employment in targeted food-insecure areas; and (b) restore access to social and economic services following natural disasters in targeted communities. | IL | New | 12 |
| The project's revised development objective is to assist war-affected or otherwise vulnerable communities to restore infrastructure and services, build local capacity for collective action, and assist vulnerable households to access temporary employment opportunities, with priority given to areas not previously serviced by the government, newly accessible areas (those that were under rebel control until January 2002), food insecure areas, and the most vulnerable population groups within those areas. | IL | AF | 4 |

| Approval FY | Project | | Economy | Region | Sector Board |
|---|---|---|---|---|---|
| | ID | Name | | | |
| 2009 | P113156 | Ethiopia Fertilizer Support Project | Ethiopia | Africa | Agriculture and Rural Development |
| 2009 | P113199 | Food Crisis Response Project | Afghanistan | South Asia | Agriculture and Rural Development |
| 2009 | P113218 | SO Rapid Response Rehab of Rural Livel | Somalia | Africa | Agriculture and Rural Development |
| 2009 | P113219 | SL—Document Policy Lending—Food Crisis Response | Sierra Leone | Africa | Social Protection |
| 2009 | P113221 | CAR Food Response Project | Central African Republic | Africa | Agriculture and Rural Development |
| 2009 | P113222 | Emergency Food Security Support Project | Niger | Africa | Agriculture and Rural Development |

| Project Development Objectives | Lending Instrument | Financing Instrument | Loan Amount (US $ millions) |
|---|---|---|---|
| To contribute to the government's efforts to ensure an aggregate availability of supply of chemical fertilizers for the 2009–10 production season, adequate to meet smallholder farmers' priority demands. | IL | New | 250 |
| To enhance wheat and other cereal production by supporting small scale irrigation at the community level through increase of irrigated land area and capacity building of communities to implement and maintain irrigation sector sub projects that address community needs. | IL | New | 8 |
| The PDO is to increase crop and livestock production in areas affected by the food crisis. | IL | New | 7 |
| The Food Crisis Response Development Policy Grant of US$3 million is a development policy grant intended to support the Government's poverty reduction strategy by providing the authorities with needed fiscal space to partially compensate for the lost revenues resulting from the recently reduced tariffs on food and fuel imports. In particular, the size of the grant was matched against the estimated incremental cost (US$3 million) of maintaining government funded feeding programs across all ministries and agencies. | DPL | New | 3 |
| The project's overall development objectives are to: (a) provide increased food access to primary and pre-school students in targeted areas; and (b) support farmer's capacity to ensure adequate supply response for medium-term improvement in food security. | IL | New | 7 |
| The objective of the grant is to support the government efforts to mitigate the impact of food price crisis through: (a) increasing rice production with procurement and distribution of fertilizers to rice producers at affordable prices; and (b) providing technical assistance to enhance the capacity of the Food Crisis Prevention and Management Framework's (FCPMF) coordination unit and executing agencies, for the management of the project and for improved monitoring of the country's food security. | IL | New | 7 |

| Approval FY | Project | | Economy | Region | Sector Board |
|---|---|---|---|---|---|
| | ID | Name | | | |
| 2008 | P113224 | Madagascar Supplemental Fifth Poverty Reduction Support Credit (PRSC V) Grant | Madagascar | Africa | Economic Policy |
| 2009 | P113232 | Global Food Price Response Program | Rwanda | Africa | Poverty Reduction |
| 2009 | P113268 | Emergency Agricultural Productivity Support Project | Guinea | Africa | Agriculture and Rural Development |
| 2009 | P113374 | Emergency Food Security Support Project | Benin | Africa | Agriculture and Rural Development |
| 2009 | P113438 | Food Crisis Response Development Policy Grant | Burundi | Africa | Economic Policy |

| Project Development Objectives | Lending Instrument | Financing Instrument | Loan Amount (US $ millions) |
|---|---|---|---|
| The urgent need for public expenditures to respond to the food price crisis has resulted in higher financing requirements than originally anticipated. This supplemental financing grant will enable the government to continue to make progress on the reform program supported by the PRSC program, which could otherwise be jeopardized by the unanticipated gap in financing for the 2008 and 2009 budgets, including the maintenance of a stable macroeconomic framework. | DPL | SF | 10 |
| The proposed supplemental financing operation to the fourth Poverty Reduction Support Grant (PRSG-IV) will fulfill the immediate needs related to sustaining food crop production and intensification. The funds will be used to import fertilizer. In the absence of such a support, indications are that there would be a continued deterioration and decline in the trend of agricultural production and yields. The agricultural growth rate during 2007 was zero and failure to provide these inputs could lead to a contraction of the sector during the current growing season. This would further exacerbate the already high and rising trend in food prices and have dire consequences for food security, particularly for the most vulnerable groups. | DPL | SF | 10 |
| The objective of the project is to increase smallholder rice productivity on 35,000 hectares. | IL | New | 5 |
| The project objective is to increase domestic production of cereals, particularly maize and rice to mitigate the short-term impact of increasing prices on households, while strengthening the capacity of the country to cope in the medium and long-term. | IL | New | 9 |
| The specific objectives of the DPG Program—in line with those of the global food crisis response—were to assist the Government of Burundi in mitigating the impact of increased food prices on the poor and to maintain fiscal stability, thus reducing the risk of food insecurity, increased poverty and social unrest, and helping to foster social and economic development. | DPL | New | 10 |

| Approval FY | Project | | Economy | Region | Sector Board |
|---|---|---|---|---|---|
| | ID | Name | | | |
| 2009 | P113468 | Food Price Crisis Response Program | Guinea-Bissau | Africa | Agriculture and Rural Development |
| 2009 | P113492 | Philippines Global Food Crisis Response Development Policy Operations | Philippines | East Asia and the Pacific | Social Protection |
| 2009 | P113544 | Additional Financing to CDD—Food Crisis | Togo | Africa | Social Protection |
| 2009 | P113586 | Emergency Food Crisis Response Project | South Sudan | Africa | Agriculture and Rural Development |
| 2009 | P113608 | Labor Intensive Public Works—Additional Financing for Third Urban Development Project Phase II | Guinea | Africa | Urban Development |
| 2009 | P113623 | Strengthening the Management of Agriculture Public Services (GFRP) | Haiti | Latin America and the Caribbean | Agriculture and Rural Development |

| Project Development Objectives | Lending Instrument | Financing Instrument | Loan Amount (US $ millions) |
|---|---|---|---|
| The objective of this project is to improve food security for the most vulnerable population, including children, and increase smallholder rice production in project areas. | IL | AF | 5 |
| The development policy operations aims to support the government of the Philippines in addressing the challenges of high food prices in the short and medium term, particularly by supporting measures to strengthen social protection and safety nets to protect poor and vulnerable households. | DPL | New | 200 |
| The proposed additional grant would support the implementation of additional subprojects for socio-economic infrastructures and income generating activities, by financing: targeted community nutrition interventions for the most vulnerable; community based activities that increase food production and facilitate the supply of food products to markets and the population; additional training; and additional operating costs. | IL | AF | 7 |
| Increased access to food for consumption by food insecure households or groups living in six of the most distressed counties of South Sudan. | IL | New | 5.7 |
| This additional financing will provide an emergency urban labor-intensive works program aimed at increasing the purchasing power of the poorest and most vulnerable in Conakry where there is real risk of social unrest exacerbated by recent increase in food prices. Work is also targeted to areas where work under the existing project indicates that road rehabilitation and drainage clearance will have the most impact. | IL | AF | 2.5 |
| To enable the Ministry of Agriculture, Natural Resources and Rural Development (MARNDR) to prioritize and target investments according to sector policies, and improve local agriculture support services. | IL | New | 5 |

| Approval FY | Project | | Economy | Region | Sector Board |
|---|---|---|---|---|---|
| | ID | Name | | | |
| 2009 | P113625 | Guinea Food Crisis Response Development Policy Grant | Guinea | Africa | Poverty Reduction |
| 2009 | P114269 | Second Poverty Reduction Support Credit Supplemental— Food Price Crisis Response Trust Fund | Mali | Africa | Social Protection |
| 2009 | P114291 | Accelerated Food Security Project | Tanzania | Africa | Agriculture and Rural Development |
| 2010 | P114375 | Additional Financing to the Second Agricultural Technology Project (GFRP) | Nicaragua | Latin America and the Caribbean | Agriculture and Rural Development |
| 2009 | P114441 | Price Vulnerability (Food Crisis) | Nicaragua | Latin America and the Caribbean | Social Protection    · |
| 2009 | P114617 | Rice Productivity Improvement Project | Lao PDR | East Asia and the Pacific | Agriculture and Rural Development |

| Project Development Objectives | Lending Instrument | Financing Instrument | Loan Amount (US $ millions) |
|---|---|---|---|
| The development objective of the grant is to support the government's Second Poverty Reduction Strategy by providing the authorities with needed fiscal space to compensate for the lost revenues resulting from the customs duties reduction of 10 percentage points on rice imports between June and October 2008. This policy measure is expected to mitigate the impact of international rice price increases and contribute to continued basic service delivery for vulnerable groups. | DPL | New | 2.5 |
| The objective of this supplemental financing is, together with the IMF and other development partner contributions, to help the government of Mali fill an unanticipated financing gap caused by the food price crisis and, thus, maintain the course of important socioeconomic policy reforms agreed under the PRSC-II. The proposed supplemental financing is part of the Bank's short to medium-term response to the food price crisis, which also includes | DPL | SF | 5 |
| The objective of the Project is to contribute to higher food production and productivity in targeted areas by improving farmers access to critical agricultural inputs. | IL | New | 160 |
| The objective of the additional financing (AF) operation is to improve the availability, access and use of certified seed in order to increase agricultural productivity in a sustainable manner in the medium term. In addition, this objective is compatible with component III.A.1 of the Global Food Crisis Response Program (GFRP) to improve smallholder access to seed and fertilizer through investment and training to strengthen existing systems for seed and fertilizer quality control. | IL | AF | 10 |
| The proposed operation focuses on mitigating the negative nutritional impact of the food price increase on pre-school and primary school children and on promoting poor rural households' food security by increasing agricultural production. | IL | New | 7 |
| The main development objective would be to increase rice productivity and overall volume of rice production among smallholders in four selected provinces, thereby substantially increasing incomes and improving food security for small farm households. | IL | New | 3 |

| Approval FY | Project | | Economy | Region | Sector Board |
|---|---|---|---|---|---|
| | ID | Name | | | |
| 2009 | P114683 | Productive Safety Nets II (FY09) Additional Financing | Ethiopia | Africa | Social Protection |
| 2010 | P114740 | Services Support Project (Co-Financing and Restructuring) | Comoros | Africa | Social Protection |
| 2010 | P114863 | Community Nutrition Project | Lao PDR | East Asia and the Pacific | Health, Nutrition, and Population |
| 2008 | P114912 | Irrigation & Water Resources Management Supplemental | Nepal | South Asia | Agriculture and Rural Development |
| 2009 | P115873 | Additional Financing for Agricultural Sector Development Project | Tanzania | Africa | Agriculture and Rural Development |
| 2009 | P115938 | Rapid Response Child-Focused Social Cash Transfer and Nutrition Security Project | Senegal | Africa | Health, Nutrition, and Population |
| 2009 | P115952 | Additional Financing for Tanzania Second Social Action Fund (TASAF II) | Tanzania | Africa | Social Protection |

| Project Development Objectives | Lending Instrument | Financing Instrument | Loan Amount (US $ millions) |
|---|---|---|---|
| This operation seeks to ensure that the needs of vulnerable chronically food insecure households are adequately addressed. To that end, the development objective is to contribute to the Government's efforts to maintain adequate coverage of the Ethiopia Productive Safety Net Program (PSNP) in 2009, thereby ensuring that the Project Development Objectives of the PSNP can be met. | IL | AF | 25 |
| (a) Increase access to short-term employment in food-insecure areas (new objective); (b) Increase access to basic social services for poor communities (reformulated objective); and (c) Contribute to building the capacity of communities to plan their development (reformulated objective). | IL | AF | 1 |
| To improve coverage of essential maternal and child health services and improve mother and child caring practices among pregnant and lactating women and children less than two years old in the seven southern and central provinces. | IL | New | 2 |
| The project development objective of the Irrigation and Water Resource Management Project remains unchanged and is to improve irrigated agriculture productivity and management of selected irrigation schemes and enhance institutional capacity for integrated water resources management. | IL | AF | 14.3 |
| The Project has two main objectives: (a) enable farmers to have better access to, and use of, agricultural knowledge, technologies, marketing systems and infrastructure, all of which contribute to higher productivity, profitability and farm incomes; and (b) promote agricultural private investment based on an improved regulatory and policy environment. | IL | AF | 30 |
| The PDO is to reduce the risk of nutrition insecurity of vulnerable populations, in particular children under five in poor rural and urban areas by scaling up the Government Nutrition Enhancement Program and providing cash transfers to vulnerable mothers of children under five. | IL | New | 10 |
| The development objective is to improve access of beneficiary households to enhanced socioeconomic services and income generating opportunities. | IL | AF | 30 |

| Approval FY | Project | | Economy | Region | Sector Board |
|---|---|---|---|---|---|
| | ID | Name | | | |
| 2009 | P116064 | Kenya Agricultural Input Supply Program | Kenya | Africa | Agriculture and Rural Development |
| 2010 | P116301 | Additional financing for food security (GFRP) | Senegal | Africa | Agriculture and Rural Development |
| 2010 | P117203 | Smallholder Agriculture and Social Protection Support Operation | Cambodia | East Asia and the Pacific | Agriculture and Rural Development |
| 2010 | P117320 | Global Food Price Crisis Response Trust Fund—Social Safety Net Reform Project (SSNRP)—Additional Food Crisis Financing | West Bank and Gaza | Middle East and North Africa | Social Protection |
| 2010 | P118226 | Second Additional Financing to NSAP - Food Crisis Response | Sierra Leone | Africa | Social Protection |
| 2010 | P120538 | Additional Financing for the Nepal Social Safety Nets Project | Nepal | South Asia | Agriculture and Rural Development |

NOTE: DPL=Development Policy Lending, IL = Investment Lending, AF=Additional Finance, SF=Supplemental Finance.

| Project Development Objectives | Lending Instrument | Financing Instrument | Loan Amount (US $ millions) |
|---|---|---|---|
| The objective of the project is to assist the recipient to increase access to agricultural inputs and technologies among targeted smallholder farmers in the six main maize producing districts. | IL | New | 5 |
| The revised project development objective is to improve competitiveness of selected domestic supply chains, increase non-traditional agricultural exports and increase rice production in project areas. Revised project outcome indicators comprise: (a) non-traditional agricultural exports in project areas reach 12,000 tons by end of project; (b) local production of onion and banana covers 75% and 50% of domestic consumption respectively; and, (c) annual production of rice paddy increases by 52,000 tons at the end of the project. | IL | AF | 10 |
| Address the weaknesses in smallholder agricultural production and social protection systems, which have come to light during the food price crisis. | DPL | New | 5 |
| The objectives of the project are to: (a) mitigate the impact of the continued socio-economic crisis on a subset of the poorest and most vulnerable households; and (b) strengthen MOSA's institutional capacity to manage cash transfer programs. | IL | AF | 3.4 |
| The project development objective is to assist war-affected or otherwise vulnerable communities to restore infrastructure and services, build local capacity for collective action, and assist vulnerable households to access temporary employment opportunities, with priority given to areas not previously serviced by the government, newly accessible areas (those which were under rebel control until January 2002), food insecure areas, and the most vulnerable population groups within those areas. | IL | AF | 3 |
| The project development objective of the restructured Social Safety Nets Project is to improve access to nutritious food for highly food insecure households in the short term and to create opportunities for improved agriculture production in food insecure districts. | IL | AF | 47.8 |

TABLE D.3  Summary Statistics for Timeliness of GFRP Project Preparation and Effectiveness (As of June 14, 2012)

| Economy | Project ID[a] | Time Between Concept Note and Approval (Days) | Time Between Approval and Effectiveness (Days) | Length of Closing Date Extension (Months) |
|---|---|---|---|---|
| Afghanistan | P113199 | 42 | 37 | 0 |
| Bangladesh | P112761 | 54 | 23 | 0 |
| Benin | P113374 | 101 | 11 | 8 |
| Burundi | P113438 | 19 | 6 | 0 |
| Cambodia | P117203 | 206 | 98 | 0 |
| Central African Republic | P113221 | 21 | 28 | 28 |
| Comoros | P114740 | 192 | 155 | 0 |
| Djibouti | P112017 | 13 | 28 | 0 |
| Ethiopia | P113156 | 83 | 13 | 1 |
| Ethiopia | P114683 | 83 | 2 | 0 |
| Guinea | P113625 | 78 | 15 | 0 |
| Guinea | P113608 | 78 | 15 | 27 |
| Guinea | P113268 | 78 | 15 | 27 |
| Guinea-Bissau | P113468 | 59 | 18 | 6 |
| Haiti | P113623 | 164 | 110 | 12 |
| Haiti | P112133 | 16 | 29 | 0 |
| Honduras | P112023 | 35 | 161 | 0 |
| Kenya | P116064 | 71 | 9 | 0 |
| Kenya | P111545 | 351 | 94 | 0 |
| Kyrgyz Republic | P112186 | 24 | 0 | 0 |
| Kyrgyz Republic | P112142 | 20 | 49 | 24 |

| Economy | Project ID[a] | Time Between Concept Note and Approval (Days) | Time Between Approval and Effectiveness (Days) | Length of Closing Date Extension (Months) |
|---|---|---|---|---|
| Lao PDR | P114617 | 103 | 48 | 0 |
| Lao PDR | P114863 | 406 | 37 | 0 |
| Liberia | P112084 | 14 | 23 | 0 |
| Liberia | P112083 | 14 | 23 | 0 |
| Liberia | P112107 | 24 | 56 | 0 |
| Madagascar | P113134 | 103 | 83 | 18 |
| Madagascar | P113224 | 23 | 1 | 0 |
| Mali | P114269 | 71 | 0 | 0 |
| Moldova | P112908 | 30 | 20 | 24 |
| Mozambique | P107313 | 124 | 30 | 0 |
| Nepal | P120538 | 77 | 133 | 0 |
| Nepal | P114912 | 63 | 33 | 0 |
| Nepal | P113002 | 63 | 120 | 24 |
| Nicaragua | P114375 | 557 | 354 | 0 |
| Nicaragua | P114441 | 98 | 114 | 4 |
| Niger | P113222 | 62 | 21 | 6 |
| Philippines | P113492 | 104 | 75 | 18 |
| Rwanda | P113232 | 19 | 1 | 0 |
| Senegal | P115938 | 85 | 128 | 0 |
| Senegal | P116301 | 370 | 121 | 0 |
| Sierra Leone | P113219 | 43 | 12 | 0 |
| Sierra Leone | P113141 | 20 | 5 | 12 |

| Economy | Project ID[a] | Time Between Concept Note and Approval (Days) | Time Between Approval and Effectiveness (Days) | Length of Closing Date Extension (Months) |
|---|---|---|---|---|
| Sierra Leone | P118226 | 65 | 7 | 0 |
| Somalia | P113218 | 39 | 13 | 0 |
| South Sudan | P113586 | 92 | 24 | 0 |
| Tajikistan | P112136 | 22 | 34 | 33.1 |
| Tajikistan | P112157 | 22 | 34 | 36 |
| Tanzania | P114291 | 174 | 78 | 0 |
| Tanzania | P115873 | 174 | 72 | 0 |
| Tanzania | P115952 | 174 | 0 | 0 |
| Togo | P113544 | 85 | 6 | 0 |
| West Bank and Gaza | P113117 | 69 | 0 | 3.9 |
| West Bank and Gaza | P117320 | 176 | 40 | 0 |
| Republic of Yemen | P112345 | 16 | 5 | 0 |
| Summary Statistics | Minimum | 13 | 0 | |
| | Maximum | 557 | 354 | |
| | Median | 71 | 28 | |
| | Mean | 97.6 | 48.5 | |
| | Standard Deviation | 106.9 | 60.9 | |
| | Coefficient of Variation | 1.1 | 1.26 | |

SOURCE: IEG Portfolio Review.
NOTE: a. Project names are in Appendix Table D.2.

TABLE D.4 Median Preparation Time (Days) for GFRP Projects, by Objectives and Regions (FY2008–11)

| World Bank Region | Objectives | | | | |
|---|---|---|---|---|---|
| | Food Price Policy Macroeconomic Stability | Social Protection | Food Supply | Mixed | Median |
| Africa (n=32) | 75 | 83 | 81 | 59 | 78 |
| Middle East and North Africa (n=4) | | 69 | | 13 | 43 |
| Europe and Central Asia (n=5) | | 22 | 22 | 24 | 22 |
| Latin American and Caribbean (n=5) | 35 | | 361 | 57 | 98 |
| South Asia (n=5) | | | 53 | 63 | 63 |
| East Asia (n=4) | | 406 | 103 | 155 | 155 |
| MEDIAN | 71 | 74 | 81 | 61 | 71 |

SOURCE: IEG Portfolio Review.

TABLE D.5 Median Time (Days) between Approval and Effectiveness for GFRP Projects, by Objectives and Regions (FY2008–11)

| World Bank Region | Objectives | | | | |
| --- | --- | --- | --- | --- | --- |
| | Food Price Policy Macroeconomic Stability | Social Protection | Food Supply | Mixed | Median |
| Africa (n=32) | 8 | 30 | 14 | 12 | 15 |
| Middle East and North Africa (n=4) | | 5 | | 28 | 17 |
| Europe and Central Asia (n=5) | | 34 | 34 | 0 | 34 |
| Latin American and Caribbean (n=5) | 161 | | 232 | 72 | 114 |
| South Asia (n=5) | | | 35 | 120 | 37 |
| East Asia (n=4) | | 37 | 48 | 87 | 62 |
| MEDIAN | 15 | 32 | 34 | 24 | 28 |

SOURCE: IEG Portfolio Review.

**TABLE D.6** IEG Ratings of Closed GFRP Operations (As of January 26, 2013)

| Index | Country | Project ID | Project Name |
|---|---|---|---|
| 1 | Afghanistan | P113199 | Food Crisis Response Project |
| 2 | Bangladesh | P112761 | Bangladesh Food Crisis Development Support Credit |
| 3 | Burundi | P113438 | Food Crisis Response Development Policy Grant |
| 4 | Cambodia | P117203 | Small Holder Agriculture and Social Protection Support Operation |
| 5 | Comoros | P114740 | Services Support Credit (parent P084315) |
| 6 | Djibouti | P112017 | Food Crisis Response Development Policy Grant |
| 7 | Ethiopia | P113156 | Ethiopia Global Food Crisis Response Pro |
| 8 | Ethiopia | P114683 | Productive Safety Nets II (FY09) Additional Financing (parent P098093) |
| 9 | Guinea | P113625 | Guinea Food Crisis Response Development Policy Grant |
| 10 | Haiti | P112133 | HT—Suppl. EGRO-II DP Grant (parent P100564) |
| 11 | Honduras | P112023 | Food Prices Crisis Supplemental to HN First Prog Fin Sec Dev Pol Credit (parent P083311) |
| 12 | Liberia | P112107 | Liberia Emergency Food Support for Vulnerable Women and Children |
| 13 | Kenya | P116064 | Kenya Agricultural Input Supply Program |
| 14 | Mali | P114269 | PRSC-II Supplemental—FPCR TF (parent P103466) |
| 15 | Mozambique | P107313 | Fifth Poverty Reduction Support Credit (PRSC5) |

| Total Loan Amount (US$ Millions) | Type of Operation | Outcome[a] | Bank Performance | | | Monitoring and Evaluation |
|---|---|---|---|---|---|---|
| | | | Quality at Entry | Quality of Supervision | Overall[b] | |
| 8 | IL | MS | MS | MS | MS | Modest |
| 130 | DPL | S | S | S | S | Modest |
| 10 | DPL | U | MU | S | MU | Modest |
| 5 | DPL | MS | MS | S | MS | Modest |
| 1 | IL—AF | S | S | S | S | Substantial |
| 5 | DPL | U | MS | MS | MS | Negligible |
| 250 | IL | MU | MU | MU | MU | Substantial |
| 25 | IL—AF | S | S | HS | S | Substantial |
| 2.5 | DPL | MU | MS | MU | MU | Negligible |
| 10 | DPL—SF | MU | MU | MU | MU | Modest |
| 10 | DPL—SF | MU | S | S | S | Modest |
| 4 | IL | MU | MS | MU | MU | Negligible |
| 5 | IL | MS | S | MS | MS | Modest |
| 5 | DPL—SF | U | MU | MU | MU | Negligible |
| 20 | DPL | S | S | S | S | Modest |

| Index | Country | Project ID | Project Name |
|---|---|---|---|
| 16 | Niger | P113222 | Emergency Food Security Support Project |
| 17 | Nicaragua | P114441 | Price Vulnerability (Food Crisis) |
| 18 | Philippines | P113492 | Philippines GFRP DPO |
| 19 | Sierra Leone | P113219 | SL-Document Policy Lending-Food Crisis Response |
| 20 | Somalia | P113218 | SO Rapid Response Rehab of Rural Livel |
| 21 | Republic of Yemen | P112345 | Additional Financing for Third Social Fund for Development Project (parent P082498) |

NOTE: DPL=Development Policy Lending, IL=Investment Lending, AF=Additional Finance, SF=Supplemental Finance, HS=Highly Satisfactory, S=Satisfactory, MS=Moderately Satisfactory, MU=Moderately Unsatisfactory, U=Unsatisfactory, HU=Highly Unsatisfactory.

a. Projects are rated based on IEG's independent reviews of the Implementation Completion and Results Reports (ICR Reviews), unless marked with an asterisk, in which case they are based on an IEG field assessment (Project Performance Assessment Report).

b. According to the OPCS/IEG Harmonized Evaluation Criteria, when the ratings for quality at entry and quality of supervision are both in the satisfactory range or both in the unsatisfactory range, overall Bank Performance is the lower of the two. When one of the two criteria is in the satisfactory range and the other in the unsatisfactory range, the Outcome rating determines whether overall Bank Performance is in the satisfactory or unsatisfactory range.

| Total Loan Amount (US$ Millions) | Type of Operation | Outcome[a] | Bank Performance | | | Monitoring and Evaluation |
| --- | --- | --- | --- | --- | --- | --- |
| | | | Quality at Entry | Quality of Supervision | Overall[b] | |
| 7 | IL | S | S | S | S | High |
| 7 | IL | MS | MS | S | MS | Substantial |
| 200 | DPL | HS | HS | S | S | Substantial |
| 3 | DPL | MS | MU | MS | MS | Modest |
| 7 | IL | MS | S | S | S | Substantial |
| 10 | IL—AF | S | MS | S | MS | Substantial |

# Appendix E
# Vulnerability Analysis

For the vulnerability analysis, three characteristics of vulnerability to adverse outcomes of food price crises were defined:

- The country's exposure to global food price spikes related to its food import or export status

- Government's capacity to respond to high food prices in terms of fiscal space, foreign exchange reserves, and social safety nets

- The magnitude of country's vulnerable population—the extent of poverty.

For each vulnerability characteristic one or more indicators were used:

- Exposure to adverse international food prices (net cereal imports as a percent of total cereal consumption)

- Government capacity to respond (total foreign exchange reserves including gold as a percent of imports of goods and services; general government revenue minus total expenditure as percent of GDP; and the SSN indicator in the CPIA in 2008)

- Extent of existing poverty (poverty headcount based on $1.25 a day as of population based on closest pre-2009 observation).

These characteristics were combined into a vulnerability index.[1] In the absence of an objective foundation to attach weights to the five indicators used in this analysis, the composite vulnerability index is a simple average of the standardized sub indicators (each of which has been converted to an index in the range 1–100 prior to aggregation). The composite index itself has been standardized to the range 1–100. The indices calculated were confined to the group of countries that were generally eligible to receive funding from the World Bank Group, namely to lower income and lower middle-income countries. A list of countries and their vulnerability index is provided in Table 3.E.1).

Similar indices or rankings have been produced by other organizations and scholars seeking to classify countries according to their vulnerability to the adverse effects of the food crisis. A study by de Janry and Sadoulet (2008)[2] assigned countries to three classes of vulnerability: most vulnerable, highly vulnerable, and somewhat vulnerable on the basis of three indicators: high food independency at the average household level, high food import burdens, and low per capita gross national income. The WFP compiled a list of countries vulnerable to the food crisis based on a hypothesis that the countries likely to exhibit high levels of food insecurity would be those that rely heavily on imported food and fuel commodities, have relatively large urban populations, are experiencing high inflationary pressures, and have populations that spend a significant proportions of their income on food.[3] Finally, the Economist Intelligence Unit recently developed a global food security index[4] based on 25 quantitative and qualitative indicators covering general themes "affordability," "availability" and "quality and safety."

To assess the extent to which GFRP funding was directed to vulnerable countries, this evaluation used the composite index constructed for the present evaluation. In order to create comparable classification with the other indexes, the following procedure was used: the group of 87 low-income and lower-middle-income countries for which a composite index was calculated were arranged in ascending order of their index value, and then divided into tesciles, each consisting of 29 countries (see Table 3.C.1). The bottom group was defined as "most vulnerable," the middle tercile as "vulnerable," and the top tercile as "less vulnerable." The resultant grouping of countries is presented in Appendix Table 3.C.2. A similar procedure was applied to the Global Food Security Index (GFSI), except that GFSI ratings were available for only 50 of the low-income and lower-middle-income countries, so terciles are composed of smaller number of countries.

---

### Endnotes

[1] As with all indices, there is some degree of arbitrariness in the choice of the indicators. For example, the World Bank—IMF "Global Monitoring Report 2012" tracks countries' vulnerability to global food price shocks using just two indicators: share of cereal imports in domestic consumption of cereal and share of food expenditures in households' total expenditure (national average).

[2] de Janvry, A. and E. Sadoulet, " The Global Food Crisis: Identification of the Vulnerable and Policy Responses," Agriculture and Resource Economics Update 12(2) 12–18, 2008, accessed at http://gianni.ucop.edu/media/are-update/files/articles/v12n2.pdf.

[3] Sanago I, and J. Luma. " Assessments of the impacts of global economic crisis on household food security: innovative approaches, lessons and challenges." Chapter 16 (pp259–273) in Omamo, S. W., U. Gentilini and S. Sanstrom (eds) Revolution: From Food Aid to Food Assistance—Innovations in Overcoming Hunger, World Food Program, Rome, 2010.

[4] http://foodsecurityindex.eiu.com/.

Composite Vulnerability Index for 87 Countries

| Country | Vulnerability Index | Country | Vulnerability Index |
|---|---|---|---|
| Somalia | 0 | Madagascar | 0.365142729 |
| South Sudan | 0 | Cote d'Ivoire | 0.37904303 |
| Eritrea | 0 | Benin | 0.381748849 |
| Vanuatu | 0.151045753 | Fiji | 0.388581659 |
| Liberia | 0.168142529 | Guinea-Bissau | 0.388880035 |
| Zimbabwe | 0.207454493 | Senegal | 0.389638832 |
| Sudan | 0.208165732 | Togo | 0.390663177 |
| Mauritania | 0.236583494 | Gambia, The | 0.393613046 |
| Papua New Guinea | 0.246626127 | Congo, Rep. | 0.393676661 |
| Central African Republic | 0.247763587 | Comoros | 0.405156553 |
| Solomon Islands | 0.259618688 | Angola | 0.422864037 |
| Mozambique | 0.274451052 | Cambodia | 0.423272012 |
| Congo, Dem. Rep. | 0.275618473 | Afghanistan | 0.424507282 |
| Swaziland | 0.279757092 | Bangladesh | 0.428803723 |
| Haiti | 0.307671854 | Yemen, Rep. | 0.429994386 |
| Belize | 0.320242595 | Nepal | 0.43097081 |
| Tanzania | 0.323264725 | Guinea | 0.432559376 |
| Chad | 0.329816852 | Uganda | 0.434018843 |
| Malawi | 0.338040263 | Kiribati | 0.443884034 |
| Djibouti | 0.345110199 | Guyana | 0.446699008 |
| Lao PDR | 0.346233488 | Philippines | 0.465871028 |
| Burundi | 0.356312061 | Pakistan | 0.47980648 |
| Rwanda | 0.360932547 | Timor-Leste | 0.484972158 |

| Country | Vulnerability Index | Country | Vulnerability Index |
|---|---|---|---|
| Kenya | 0.491799712 | Micronesia, Fed. Sts. | 0.639344774 |
| Zambia | 0.491928017 | Tajikistan | 0.641642934 |
| Ethiopia | 0.493880559 | Sao Tome and Principe | 0.655155989 |
| Sierra Leone | 0.497337912 | Kosovo | 0.655948884 |
| Samoa | 0.508880575 | Armenia | 0.656459978 |
| Tonga | 0.519928271 | Georgia | 0.659388619 |
| Sri Lanka | 0.523656496 | Kyrgyz Republic | 0.66176646 |
| Bhutan | 0.540056153 | Guatemala | 0.663953811 |
| Cape Verde | 0.544936491 | Bolivia | 0.670456177 |
| Burkina Faso | 0.544982394 | Turkmenistan | 0.673854654 |
| Moldova | 0.545576118 | Ghana | 0.68071523 |
| India | 0.550250745 | El Salvador | 0.701231089 |
| Indonesia | 0.560848315 | Paraguay | 0.702531568 |
| Nigeria | 0.561719176 | Uzbekistan | 0.709699495 |
| Vietnam | 0.576730292 | Ukraine | 0.718194718 |
| Egypt, Arab Rep. | 0.580438306 | Cameroon | 0.72508194 |
| Mali | 0.586907455 | Niger | 0.7668097 |
| Mongolia | 0.598881621 | | |
| Honduras | 0.606511201 | | |
| Nicaragua | 0.620899906 | | |
| Syrian Arab Republic | 0.621722332 | | |
| Morocco | 0.637244281 | | |
| Lesotho | 0.638089387 | | |
| Marshall Islands | 0.639344774 | | |

TABLE E.2 GFRP Economies According to the Number of Vulnerability Indexes
(IEG/WFP/de Janvry/GFSI Index)

| Economy Received GFRP loans | Number of Operations | IEG Index[a] | WFP Index[b] | de Janvry Index[c] | GFSI Index[d] | Total Appearances[f] |
|---|---|---|---|---|---|---|
| Bangladesh | 1 | 1 | 1 | 1 | 1 | 4 |
| Benin | 1 | 1 | 1 | 1 | 1 | 4 |
| Burundi | 1 | 1 | 1 | 1 | 1 | 4 |
| Ethiopia | 2 | 1 | 1 | 1 | 1 | 4 |
| Guinea | 3 | 1 | 1 | 1 | 1 | 4 |
| Haiti | 2 | 1 | 1 | 1 | 1 | 4 |
| Kenya | 2 | 1 | 1 | 1 | 1 | 4 |
| Madagascar | 2 | 1 | 1 | 1 | 1 | 4 |
| Mozambique | 1 | 1 | 1 | 1 | 1 | 4 |
| Rwanda | 1 | 1 | 1 | 1 | 1 | 4 |
| Senegal | 2 | 1 | 1 | 1 | 1 | 4 |
| Sierra Leone | 3 | 1 | 1 | 1 | 1 | 4 |
| Tanzania | 3 | 1 | 1 | 1 | 1 | 4 |
| Togo | 1 | 1 | 1 | 1 | 1 | 4 |
| Republic of Yemen[e] | 1 | 1 | 1 | 1 | 1 | 4 |
| Afghanistan | 1 | 1 | 1 | 1 | 0 | 3 |
| Central African Republic | 1 | 1 | 1 | 1 | 0 | 3 |
| Guinea-Bissau | 1 | 1 | 1 | 1 | 0 | 3 |
| Nepal | 3 | 1 | 1 | 0 | 1 | 3 |
| Niger | 1 | 0 | 1 | 1 | 1 | 3 |
| Sudan | 1 | 1 | 0 | 1 | 1 | 3 |

| Economy Received GFRP loans | Number of Operations | IEG Index[a] | WFP Index[b] | de Janvry Index[c] | GFSI Index[d] | Total Appearances[f] |
|---|---|---|---|---|---|---|
| Tajikistan | 2 | 0 | 1 | 1 | 1 | 3 |
| Cambodia | 1 | 1 | 0 | 0 | 1 | 2 |
| Comoros | 1 | 1 | 1 | 0 | 0 | 2 |
| Liberia | 3 | 1 | 0 | 1 | 0 | 2 |
| Mali | 1 | 0 | 0 | 1 | 1 | 2 |
| Somalia | 1 | 1 | 1 | 0 | 0 | 2 |
| Djibouti | 1 | 1 | 0 | 0 | 0 | 1 |
| Kyrgyz Republic | 2 | 0 | 0 | 1 | 0 | 1 |
| Lao PDR | 2 | 1 | 0 | 0 | 0 | 1 |
| Moldova | 1 | 1 | 0 | 0 | 0 | 1 |
| Philippines | 1 | 1 | 0 | 0 | 0 | 1 |
| West Bank and Gaza | 2 | 0 | 1 | 0 | 0 | 1 |
| Honduras | 1 | 0 | 0 | 0 | 0 | 0 |
| Nicaragua | 2 | 0 | 0 | 0 | 0 | 0 |
| TOTAL[g] | 55 | 28 | 24 | 24 | 21 | |

NOTE: a. "On IEG Index" indicates whether a GFRP recipient is classified as "most vulnerable" or "vulnerable" in the analysis.
b. "WFP Index" refers to countries listed in Table 16.1 in "Assessments of the impacts of global economic crises on household food security: innovative approaches, lessons and challenges," Issa Sanogo and Joyce K. Luma WFP.
c. "On de Janvry Index" indicates whether a GFRP recipient is classified as "most vulnerable" or "highly vulnerable" in "The Global Food Crisis: Identification of the Vulnerable and Policy Responses," Alain de Janvry and Elisabeth Sadoulet.
d. "GFSI Index" used in this table is not the full published index, but only includes low- and lower middle-income countries (except Myanmar which is not included in our index). These countries (totally 49), ranked from low to high based on their GFSI index, are classified into three equal-size groups (most vulnerable, vulnerable, and less vulnerable). "On GFSI Index" indicates whether a GFRP recipient belongs to "most vulnerable" or "vulnerable" group.
e. Republic of Yemen is shown as both "highly vulnerable" and "somewhat vulnerable" in Table 1 in de Janvry (2008). We regard the Republic of Yemen as "highly vulnerable" when constructing this table.
f. "Total Number of Appearance" indicates how many times a country appears in the above four indexes.
g. The last row "Total" tells the number of GFRP economies that show up in each index.

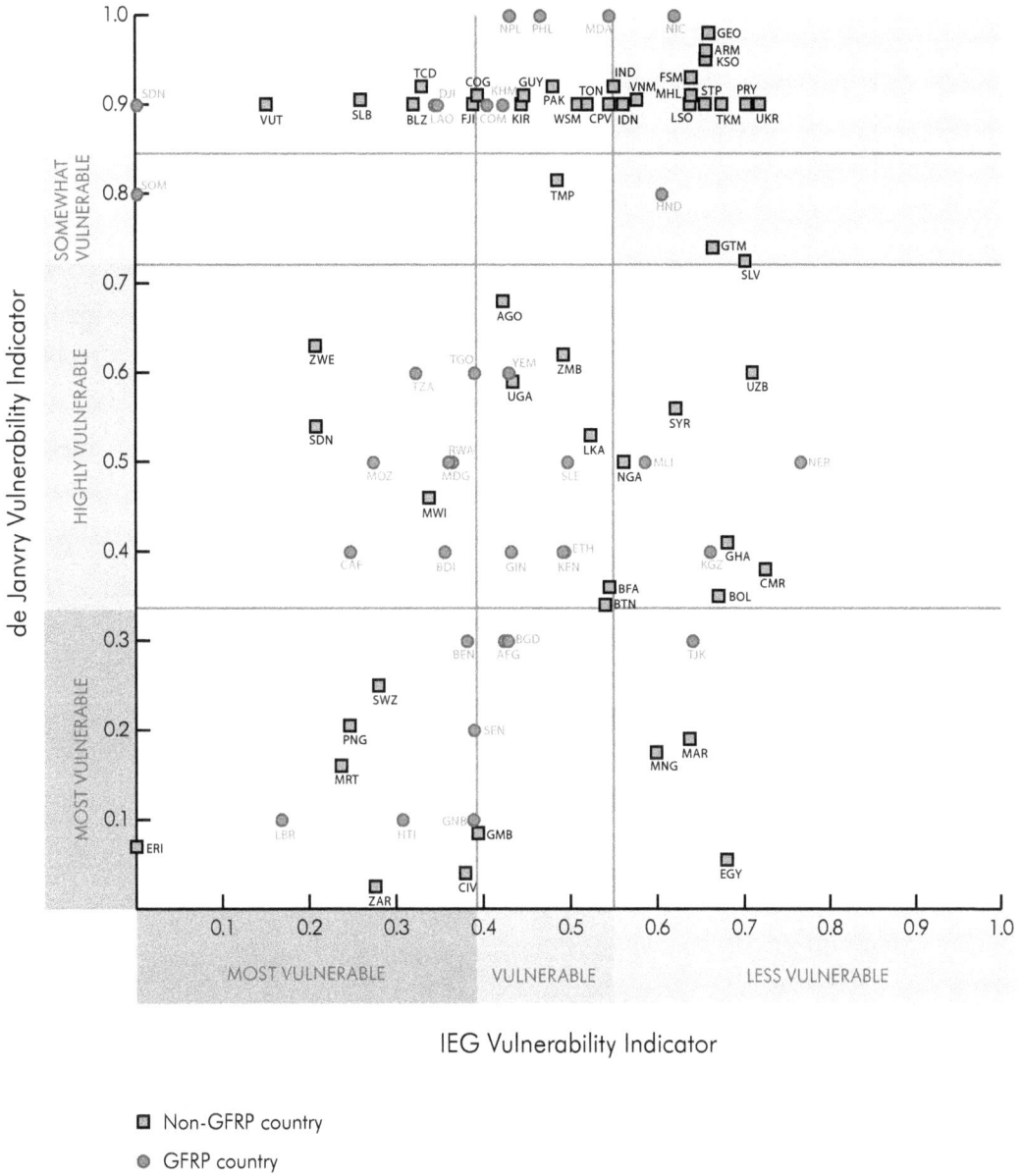

FIGURE E.1 Country Vulnerability: IEG and de Janvry Indicators Compared

IEG Vulnerability Indicator

☐ Non-GFRP country
● GFRP country

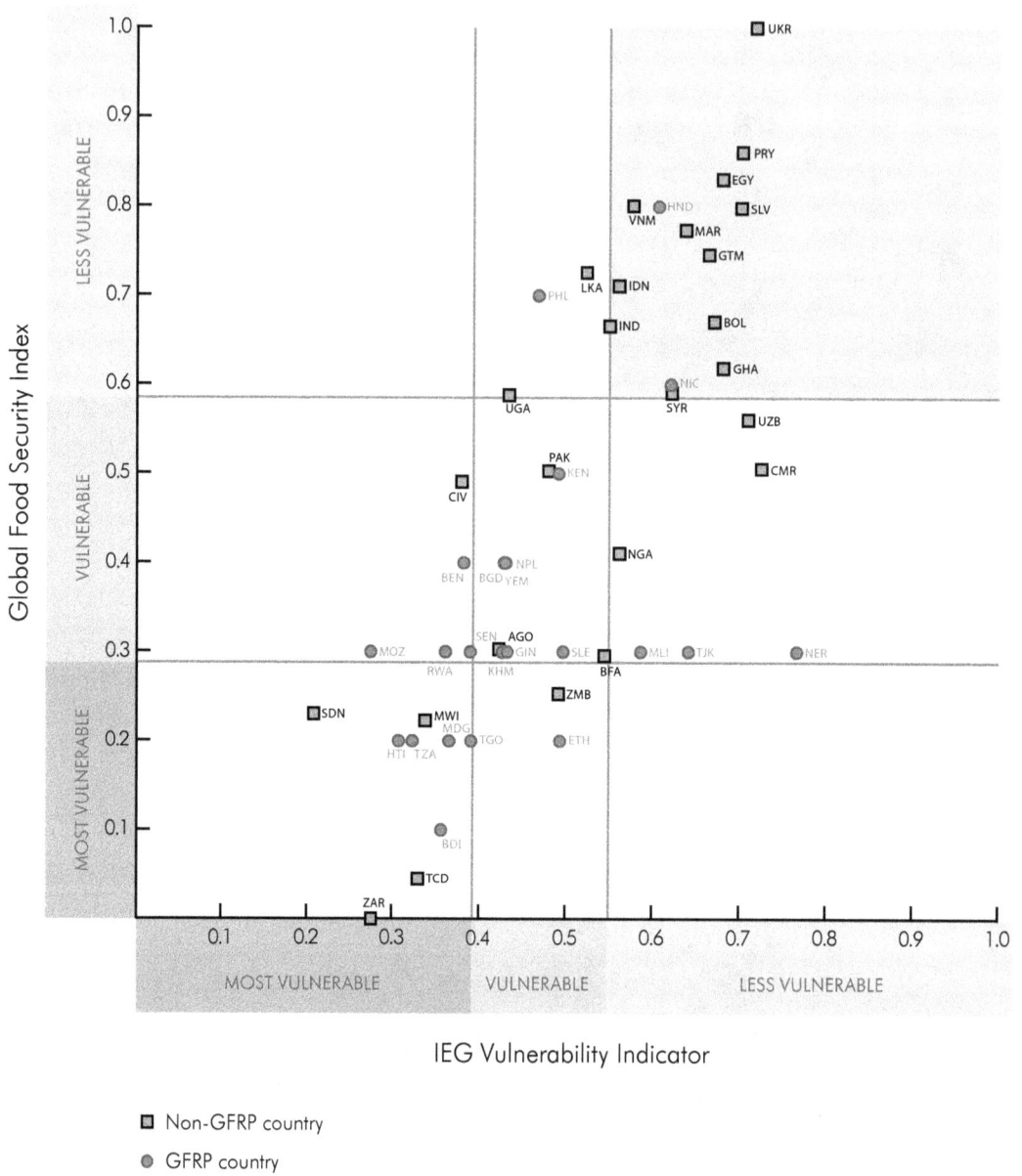

**FIGURE E.2** Country Vulnerability: IEG Indicator and Global Food Security Index Compared

IEG Vulnerability Indicator

□ Non-GFRP country

● GFRP country

# Appendix F
## Review of the Agriculture Lending Portfolio

All IBRD and IDA projects were coded by the originating unit using a standard set of codes for sectors and themes. Project coding was verified during the project approval process by the Bank teams. To identify the agriculture portfolio, IEG first searched all active and closed investment and policy operations approved between FY2006 and FY2011. Then, consistent with the Agriculture Action Plan,[1] projects were retained if they had any of the agriculture sector codes[2] or they were managed under the oversight of the Agriculture and Rural Development Sector Board. In addition, only projects carried out by the Bank, with product line defined as IBRD/IDA/Special Financing, were included in the final portfolio. Thus, 28 recipient-executed projects and 102 Global Environment Facility projects were excluded.

Distribution of identified projects by approval year is shown in the following table. After a steady increase from 80 operations in FY2006, number of agriculture projects reached its peak at 103 in FY2009, but a significant decline followed with 76 operations approved in FY2011, even lower than that in FY2006. This trend is consistent for projects under the ARD Sector Board and other Sector Boards.

Consistent with the Agriculture Action Plan, IEG used the following criteria to calculate the commitments:

- For projects with any of the agricultural sector codes but not under the oversight of the Agriculture and Rural Development Sector Board, the Agricultural amount corresponded to the share of the sector code(s) multiplied by the full dollar value of the projects.

- For projects under the oversight of the Agricultural and Rural Development Sector Board, the full dollar value of the project was counted as Agricultural amount.

On average, 80 percent of the agricultural commitments were managed under the Agriculture and Rural Development Sector Board, the remaining 20 percent under other Sector Boards. Consistent with the trend in number of operations, the agricultural commitments peaked at nearly $5.3 billion in FY2009, and more than doubled compared to FY2008 ($2.6 billion). Commitments under other sector boards contributed a significant portion of this increase, with the amount almost tripled from $0.6 billion to $1.6 billion and thus accounting for slightly more than 30 percent of the total agricultural commitments in FY2009. After that, total annual lending fell from $5.3 billion to $3.6 billion, and 65 percent of the decline were due to the drop in other sector boards' commitments.

In terms of sector distribution, commitments in agriculture, fishing, and forestry[3] accounted for, on average, 60 percent of the total IBRD/IDA agricultural lending, followed by other agriculture related investments[4] (30 percent) and agriculture markets, trade, and agro-industry[5] (10 percent). However, commitments in agriculture markets, trade, and agro-industry experienced a significant increase during this period, rising from only 3 percent in FY2007 to 21 percent of total IBRD/IDA agricultural lending in FY2011.

## Endnotes

[1] World Bank Group Agriculture Action Plan 2013–2015, July 2012.

[2] Agriculture sector codes include AB (Agriculture Extension and Research), AH (Crops), AI (Irrigations and Drainage), AJ (Animal Production and Fishing), AT (Forestry), AZ (General Agriculture), YA (Agro-industry, Marketing and Trade), BL (Public Administration-Agriculture, Fishing and Forestry). OPCS merged sector codes YA (Agricultural marketing and trade) and YB (Agro-industry), thus some projects are included if they have sector code YB.

[3] Commitments in this group correspond to the proportional dollar value allocated to agriculture sectors AB, AH, AI, AJ, AT, and AZ.

[4] This includes investments under the oversight of the Agriculture and Rural Development Sector Board other than those coded under agriculture, fishing and forestry; and agriculture markets, trade, and agro-industry and public-administration-agriculture.

[5] Commitments in this group correspond to the proportional dollar value allocated to agriculture sectors YA, YB, and BL.

Distribution of IBRD/IDA Agricultural Projects by Approval Year

| Approval Year | Managed by the ARD Sector | Managed by Other Sectors | Total |
|---|---|---|---|
| FY06 | 53 | 27 | 80 |
| FY07 | 54 | 25 | 79 |
| FY08 | 53 | 38 | 91 |
| FY09 | 61 | 42 | 103 |
| FY10 | 53 | 36 | 89 |
| FY11 | 48 | 28 | 76 |
| TOTAL | 322 | 196 | 518 |

TABLE F.2 Distribution of IBRD/IDA Agricultural Lending by Approval Year

| Approval Year | Managed by the ARD Sector | | Managed by Other Sectors | | Total Lending (US$ Millions) |
|---|---|---|---|---|---|
| | Commitments (US$ Millions) | Percentage of Total Lending | Commitments (US$ Millions) | Percentage of Total Lending | |
| FY06 | 2,461 | 86 | 412 | 14 | 2,873 |
| FY07 | 3,022 | 92 | 273 | 8 | 3,295 |
| FY08 | 2,071 | 79 | 554 | 21 | 2,625 |
| FY09 | 3,632 | 69 | 1,638 | 31 | 5,269 |
| FY10 | 3,252 | 79 | 886 | 21 | 4,138 |
| FY11 | 3,046 | 85 | 545 | 15 | 3,591 |
| TOTAL | 17,483 | 80 | 4,307 | 20 | 21,791 |
| AVERAGE | 2,914 | 80 | 718 | 20 | 3,632 |

TABLE F.3  Distribution of IBRD/IDA Agricultural Lending by Sub-Sector

| | FY06 | FY07 | FY08 | FY09 | FY10 | FY11 | Average |
|---|---|---|---|---|---|---|---|
| IBRD/IDA (by Sub-Sector) | 2,873 | 3,295 | 2,625 | 5,269 | 4,138 | 3,591 | 3,632 |
| Agricultural Production and Markets | 1,935 | 1,809 | 1,525 | 4,142 | 3,021 | 2,882 | 2,552 |
| Of Which, Agriculture, Fishing and Forestry | 1,754 | 1,719 | 1,369 | 3,469 | 2,618 | 2,129 | 2,176 |
| Percentage of Total Agricultural Production and Markets Lending | 91 | 95 | 90 | 84 | 87 | 74 | 85 |
| Percentage of Total IBRD/IDA Lending | 61 | 52 | 52 | 66 | 63 | 59 | 60 |
| Of Which, Agriculture Markets, Trade, Agro-Industry | 181 | 90 | 156 | 672 | 403 | 753 | 376 |
| Percentage of Total Agricultural Production and Markets Lending | 9 | 5 | 10 | 16 | 13 | 26 | 15 |
| Percentage of Total IBRD/IDA Lending | 6 | 3 | 6 | 13 | 10 | 21 | 10 |
| Other Agriculture Related Investments | 938 | 1,486 | 1,100 | 1,128 | 1,117 | 708 | 1,080 |
| Percentage of Total IBRD/IDA Lending | 33 | 45 | 42 | 21 | 27 | 20 | 30 |

TABLE F.4 Ratings of Development Outcomes in Pre- and Post-Crisis Period

| Project Type/ Period | Number of Projects | Outcome Rating | | | | | |
|---|---|---|---|---|---|---|---|
| | | Highly Satisfactory/ Satisfactory | | Moderately Satisfactory | | Moderately Satisfactory or Better | |
| | | Number | Percentage | Number | Percentage | Number | Percentage |
| Pre-Crisis 2006–08 | | | | | | | |
| ARD Sector | 89 | 42 | 47 | 34 | 38 | 76 | 85 |
| Other Sectors | 92 | 39 | 42 | 34 | 37 | 73 | 79 |
| Total Agriculture | 181 | 81 | 45 | 68 | 38 | 149 | 82 |
| Rest of Bank Cohort | 560 | 240 | 43 | 193 | 34 | 433 | 77 |
| Post-Crisis 2009–11 | | | | | | | |
| ARD Sector | 66 | 19 | 29 | 28 | 42 | 47 | 71 |
| Other Sectors | 52 | 12 | 23 | 23 | 44 | 35 | 67 |
| Total Agriculture | 118 | 31 | 26 | 51 | 43 | 82 | 69 |
| Rest of Bank Cohort | 388 | 118 | 30 | 165 | 43 | 283 | 73 |

TABLE F.5 Ratings of Risk to Development Outcomes in Pre- and Post-Crisis Periods[a]

| Project Type/ Period | Number of Projects | Risk Rating | | | |
| | | High or Significant | | Moderate or Negligible/Low | |
| | | Number | Percentage | Number | Percentage |
| Pre-Crisis 2006–08 | | | | | |
| ARD Sector | 88 | 38 | 43 | 50 | 57 |
| Other Sectors | 91 | 33 | 36 | 58 | 64 |
| Total Agriculture | 179 | 71 | 40 | 108 | 60 |
| Rest of Bank Cohort | 551 | 170 | 31 | 381 | 69 |
| Post-Crisis 2009–11 | | | | | |
| ARD Sector | 66 | 38 | 58 | 28 | 42 |
| Other Sectors | 51 | 26 | 51 | 25 | 49 |
| Total Agriculture | 117 | 64 | 55 | 53 | 45 |
| Rest of Bank Cohort | 388 | 164 | 42 | 224 | 58 |

NOTE: a. Some of the projects completed in 2006 were rated for likelihood of sustainability rather for the risk to development outcome. The convention used in this table is that projects rated "highly likely", "likely" or "uncertain" on sustainability were classified as having "low or negligible risk to development outcomes", while those classified as "unlikely" or "highly unlikely" to be sustainable were classified as "high or significant risk to development outcomes.

TABLE F.6  Ratings of Borrower Performance in Pre- and Post-Crisis Periods

| Project Type/Period | Number of Projects | Borrower Performance Rating | | | | | |
| | | Highly Satisfactory/ Satisfactory | | Moderately Satisfactory | | Moderately Satisfactory or Better | |
| | | Number | Percentage | Number | Percentage | Number | Percentage |
| Pre-Crisis 2006–08 | | | | | | | |
| ARD Sector | 91 | 40 | 44 | 31 | 34 | 71 | 78 |
| Other Sectors | 94 | 48 | 51 | 27 | 29 | 75 | 80 |
| Total Agriculture | 185 | 88 | 48 | 58 | 31 | 146 | 79 |
| Rest of Bank Cohort | 569 | 275 | 48 | 172 | 30 | 447 | 79 |
| Post-Crisis 2009–11 | | | | | | | |
| ARD Sector | 67 | 22 | 33 | 23 | 34 | 45 | 67 |
| Other Sectors | 52 | 19 | 37 | 19 | 37 | 38 | 73 |
| Total Agriculture | 119 | 41 | 34 | 42 | 35 | 83 | 70 |
| Rest of Bank Cohort | 395 | 114 | 29 | 168 | 43 | 282 | 71 |

TABLE F.7 Ratings on Bank Quality-at-Entry in Pre- and Post-Crisis Periods

| Project Type/ Period | Number of Projects | Quality at Entry Rating | | | | | |
|---|---|---|---|---|---|---|---|
| | | Highly Satisfactory/ Satisfactory | | Moderately Satisfactory | | Moderately Satisfactory or Better | |
| | | Number | Percentage | Number | Percentage | Number | Percentage |
| Pre-Crisis 2006–08 | | | | | | | |
| ARD Sector | 89 | 32 | 36 | 24 | 27 | 56 | 63 |
| Other Sectors | 94 | 46 | 49 | 30 | 32 | 76 | 81 |
| Total Agriculture | 183 | 78 | 43 | 54 | 30 | 132 | 72 |
| Rest of Bank Cohort | 570 | 296 | 52 | 140 | 25 | 436 | 76 |
| Post-Crisis 2009–11 | | | | | | | |
| ARD Sector | 67 | 17 | 25 | 20 | 30 | 37 | 55 |
| Other Sectors | 52 | 15 | 29 | 18 | 35 | 33 | 63 |
| Total Agriculture | 119 | 32 | 27 | 38 | 32 | 70 | 59 |
| Rest of Bank Cohort | 397 | 125 | 31 | 141 | 36 | 266 | 67 |

**TABLE F.8** Ratings of Bank Quality of Supervision in Pre- and Post-Crisis Periods

| Project Type/Period | Number of Projects | Quality of Supervision Rating | | | | | |
| | | Highly Satisfactory/ Satisfactory | | Moderately Satisfactory | | Moderately Satisfactory or Better | |
| | | Number | Percentage | Number | Percentage | Number | Percentage |
| Pre-Crisis 2006–08 | | | | | | | |
| ARD Sector | 88 | 60 | 68 | 17 | 19 | 77 | 88 |
| Other Sectors | 94 | 64 | 68 | 17 | 18 | 81 | 86 |
| Total Agriculture | 182 | 124 | 68 | 34 | 19 | 158 | 87 |
| Rest of Bank Cohort | 559 | 368 | 66 | 113 | 20 | 481 | 86 |
| Post-Crisis 2009–11 | | | | | | | |
| ARD Sector | 67 | 28 | 42 | 18 | 27 | 46 | 69 |
| Other Sectors | 51 | 22 | 43 | 13 | 25 | 35 | 69 |
| Total Agriculture | 118 | 50 | 42 | 31 | 26 | 81 | 69 |
| Rest of Bank Cohort | 390 | 191 | 49 | 132 | 34 | 323 | 83 |

TABLE F.9 Selected Performance Indicators for GFRP Agricultural Projects

| Project IDᵃ | Country | Number of Farmers Reached | Number of Hectares Covered | Yield Increase |
|---|---|---|---|---|
| P113119 | Afghanistan | | | |
| P112761 | Bangladesh | | | |
| P113374 | Benin | 37,543 | | |
| P113438 | Burundi | | | |
| P117203 | Cambodia | | | |
| P113221 | Central African Republic | 13,475 (seed)<br>2,771 (animals)<br>4,000 (tools)<br>7,608 (training) | | |
| P113156 | Ethiopia | More than 3 million | | |
| P113468 | Guinea Bissau | | | |
| P113268 | Guinea | 23,835 | | |
| P113623 | Haiti | 3,000 | | |
| P116064 | Kenya | 55,135 | | |
| P112186 | Kyrgyz Republic | 66 communities (about 2,400 farmers) | | 30%–133%, median 67% |
| P114617 | Lao PDR | 4,000 | | |

| Tons of Fertilizers | Tons of Seeds | Tons of Output Gained | Comments |
|---|---|---|---|
| | | | 474 sub projects completed (minor irrigation) |
| | | | 704 new fertilizer dealers approved |
| 9,143 | 10,860 (Rice) 54,000 (Maize) | | |
| | | | |
| | | | |
| | 510,370 imported 427,000 in 1st season | | 7% increase |
| | | 9,100 | |
| 1,000 | 1,600 | | 15,290 liters of pesticides |
| | | | as of last ISR only 300 reached |
| | | | Maize produced by beneficiaries is 3%–3.5% of national supply |
| 1,254 | 790 | | |
| | | | 4,000 farmers are the planned target. Project is much delayed. |

| Project ID[a] | Country | Number of Farmers Reached | Number of Hectares Covered | Yield Increase |
|---|---|---|---|---|
| P112083 | Liberia | | | |
| P107313 | Mozambique | | | |
| P114912 | Nepal | | | |
| P120538 | Nepal | 24,200 per year | | |
| P114441 | Nicaragua | 8,810 | 17,597 | 32% |
| P114375 | Nicaragua | | | |
| P113222 | Niger | 20,784 | 9,265 | 116% |
| P116301 | Senegal | | 6,991 | |
| P113218 | Somalia | 35,000 | 16,000 | 38%–100% |
| P113586 | South Sudan | 189,000 | | |
| P112157 | Tajikistan | 94,000 | | 17½% (wheat) |
| P114291 | Tanzania | 2 million | 485,629 | 34% |
| P115873 | Tanzania | | | |

| Tons of Fertilizers | Tons of Seeds | Tons of Output Gained | Comments |
|---|---|---|---|
| | | | Project much delayed. Infrastructure improvements were supposed to benefit 150,000 farmers |
| | | | Irrigated area increase 2,062 hectares, intensive production area increase 88,000 for maize, 115,327 for rice. Agricultural budget increased from 4.79% to 6% of total |
| | | | Target was 12 400 HH, Country has 1.7 million holdings of more than 0.5 hectares |
| | Year 1 - 700 Year 2 - 1,150 | | 851 rural roads rehabilitated |
| | | | |
| | | | Project much delayed. |
| 4,000 | | | |
| | | 34,652 paddy in two seasons | |
| | | 100,000 grain 11,875 meat | |
| | | | Number of farmers who have adopted at least one improved practice |
| | | | |
| 83,052 | 113,119 | | |
| | | | This was "additional finance "and no separate performance indicators provided for GFRP Project |

| Project ID[a] | Country | Number of Farmers Reached | Number of Hectares Covered | Yield Increase |
|---|---|---|---|---|
| P115952 | Tanzania | | | |
| P113544 | Togo | | | |

NOTE: ISR=implementation status report.
a. Project names are in Appendix Table 3.B.

| Tons of Fertilizers | Tons of Seeds | Tons of Output Gained | Comments |
|---|---|---|---|
|  |  |  | Same as above |
| 209 maize<br>94 rice |  |  | Appraisal target was 14,000 farmers |

# Appendix G
# World Bank Group Agriculture Advisory Activities

## World Bank AAA

For the AAA portfolio review, IEG downloaded from an internal Bank database all the economic and sector work (ESW) and non-lending technical assistance projects approved between FY2006 and FY2011.[1] Applying the Agriculture Action Plan definition, projects with agriculture sector codes or managed under the Agriculture and Rural Development Sector Board were included in the final AAA portfolio.

Among the 891 identified activities, 58 percent were ESWs and the remaining 42 percent were non-lending technical assistance. Although ESW activities account for nearly 60 percent of the total agriculture AAA activities, its share declined from 78 percent in FY2006 to 48 percent in FY2011. On the other hand, nonlending technical assistance activities increased dramatically from 22 percent to 52 percent, and for the first time, exceeded ESW activities in FY2011. The drop in ESWs and the increase in non-lending TAs are more pronounced when FY2003–05 is considered. During this period among the 469 identified activities 71 percent are ESWs and 29 percent are non-lending TAs.

Unlike the lending portfolio, the majority of agriculture AAA activities are not managed under the Agriculture and Rural Development Sector Board. As seen in the following table, the number of activities under other sector boards is always higher than that under the ARD sector board. For the period FY2006–08, 43 percent of agriculture AAA activities were managed under the ARD sector board; this number further dropped to 35 percent during FY2009–11 period. In line with this percentage change, the number of ARD-managed activities declined steadily, while that of non-ARD managed activities experienced a moderate increase in the recent two years. Overall, the number of agriculture-related AAA activities decreased from 456 in pre-crisis period to 435 in post-crisis period.

For both ESW and non-lending technical assistance activities, the dollar amounts downloaded correspond to expenditures and not commitments. Consistent with the criteria adopted in the lending portfolio, IEG calculated the agriculture AAA expenditures in the following way:[2]

- If a project is managed under the Agriculture and Rural Development Sector Board, the whole amount delivered is counted as Agriculture AAA spending;

- If a project is managed under other Sector Boards, only the proportional amount allocated to agriculture sectors is regarded as Agriculture AAA spending.

Although overall expenditure on Agriculture AAA stagnated over 2006–2011, expenditure associated with other sector boards rose significantly in the post-crisis period, so that expenditure in FY2009–11 is 52 percent more than that of pre-crisis period (FY2006–08), which is in line with the change in the number of AAA activities managed under other sector boards. In particular, non-lending technical assistance expenditure increased more than ESW's. In contrast, ARD-managed AAA activities experienced a 17 percent drop in expenditure. The 22 percent increase in non-lending technical assistance helped mitigate the even larger decline (32 percent) in expenditure among ESW activities under the ARD sector board.

---

## Endnotes

[1] Projects approved between FY2003 and FY2005 were also downloaded to show a longer trend of Agriculture AAA commitments and number of activities, but are not used for further analysis comparing pre- and post-crisis periods.

[2] There are 14 projects (1 in FY2006–08 and 13 in FY2009–11, representing 0.2% and 3% of the total AAA activities) with missing information on total cumulative cost delivered, which create missing information on expenditures allocated to agriculture. By assuming same activity size among these missing projects, IEG manually inflated the total expenditure amounts by 0.2% for FY2006–08 and 3% for FY2009–11.

**TABLE G.1** Distribution of Agricultural Analytical and Advisory Activities by Approval Year and Product Line

| Approval Year | ESW | | Non-Lending TA | | Total |
|---|---|---|---|---|---|
| | Number of Activities | Percentage of Total Activities | Number of Activities | Percentage of Total Activities | |
| FY06 | 120 | 78 | 33 | 22 | 153 |
| FY07 | 90 | 55 | 73 | 45 | 163 |
| FY08 | 88 | 63 | 52 | 37 | 140 |
| FY09 | 68 | 52 | 64 | 48 | 132 |
| FY10 | 77 | 52 | 72 | 48 | 149 |
| FY11 | 74 | 48 | 80 | 52 | 154 |
| **TOTAL** | 517 | 58 | 374 | 42 | 891 |

NOTE: ESW = economic and sector work, TA = technical assistance.

TABLE G.2 Distribution of Agricultural Analytical and Advisory Activities by Approval Year and Sector Board

| Approval Year | Managed by the ARD Sector | | Managed by Other Sectors | | Total |
|---|---|---|---|---|---|
| | Number of Activities | Percentage of Total Activities | Number of Activities | Percentage of Total Activities | |
| FY2006 | 70 | 46 | 83 | 54 | 153 |
| FY2007 | 69 | 42 | 94 | 58 | 163 |
| FY2008 | 55 | 39 | 85 | 61 | 140 |
| FY2009 | 49 | 37 | 83 | 63 | 132 |
| FY2010 | 47 | 32 | 102 | 68 | 149 |
| FY2011 | 57 | 37 | 97 | 63 | 154 |
| FY2006–08 | 194 | 43 | 262 | 57 | 456 |
| FY2009–11 | 153 | 35 | 282 | 65 | 435 |
| PERCENTAGE CHANGE | -21 | -17 | 8 | 13 | -5 |

NOTE: ARD = agriculture and rural development.

TABLE G.3 Agricultural Analytical and Advisory Activities Expenditure by Approval Year and Sector Board (US$ Millions)

| Approval Year | ESW | | | Non-Lending TA | | | Total Expenditure | | |
|---|---|---|---|---|---|---|---|---|---|
| | Under ARD Sector | Under Other Sectors | Total | Under ARD Sector | Under Other Sectors | Total | Under ARD Sector | Under Other Sectors | Total |
| FY2006 | 14.9 | 3.4 | 18.3 | 4.5 | 1.2 | 5.7 | 19.4 | 4.6 | 24 |
| FY2007 | 8.8 | 3.5 | 12.3 | 5 | 3.8 | 8.7 | 13.7 | 7.3 | 21 |
| FY2008 | 8.2 | 4 | 12.2 | 3.5 | 3.3 | 6.8 | 11.7 | 7.3 | 19 |
| FY2009 | 8.5 | 3.5 | 12 | 4.5 | 3 | 7.5 | 13.1 | 6.5 | 19.6 |
| FY2010 | 5.6 | 6.9 | 12.5 | 4.3 | 5.4 | 9.7 | 10 | 12.2 | 22.2 |
| FY2011 | 7.5 | 5.5 | 13 | 6.9 | 4.8 | 11.7 | 14.4 | 10.3 | 24.7 |
| FY2006–08 | 31.9 | 10.9 | 42.8 | 12.9 | 8.3 | 21.2 | 44.9 | 19.2 | 64.1 |
| FY2009–11 | 21.6 | 15.9 | 37.5 | 15.8 | 13.2 | 29 | 37.4 | 29.1 | 66.5 |
| PERCENTAGE CHANGE (PRE- VS. POST-CRISIS) | -32 | 46 | -12 | 22 | 59 | 36 | -17 | 52 | 4 |

NOTE: ARD = agriculture and rural development, ESW = economic and sector work, TA = technical assistance.

TABLE G.4  IFC Advisory Services in Pre- and Post-Crisis Periods

| Region | Cost of Advisory Services (US$ Millions) | | Percentage Distribution | |
|---|---|---|---|---|
| | FY2006–08 | FY2009–11 | FY2006–08 | FY2009–11 |
| AFR | 1.9 | 1.2 | 14.9 | 6.3 |
| EAP | 4.7 | 2.3 | 36.8 | 12.6 |
| ECA | 0.6 | 4.7 | 5 | 25.2 |
| LCR | 2.5 | 2.2 | 19.7 | 11.8 |
| MNA | 2.6 | 4.2 | 20.6 | 22.9 |
| SAR | 0.4 | 3.9 | 3 | 21.2 |
| TOTAL | 12.8 | 18.5 | 100 | 100 |

SOURCE: IFC data.

# Appendix H
# Social Safety Net Portfolio Review

## The Social Safety Net Lending Portfolio

The World Bank's social safety net (SSN) portfolio is used throughout the evaluation to assess trends in lending and performance. Project variables are taken from the Bank's operational database (Business Warehouse), IEG's 2012 Evaluation of Social Safety Nets,[1] and project documents.

### IDENTIFICATION

Overall, there are 200 free-standing projects included in the Social Safety Net portfolio for this evaluation. This portfolio included: all 130 projects approved between FY2006 and FY2010 from the SSN evaluation;[2] an additional 34 IBRD/IDA safety net projects approved in FY2011; 26 projects[3] that were assigned the new theme code 91 (Global Food Crisis Response) with a social safety net component; 7 special funding projects thematically coded as 54 (Social Safety Nets); and 3 projects with the social safety net theme code and approved between FY2006 and FY2010.[4] In addition to these 200 projects, there were 23 supplemental projects approved FY2006–11 with a social safety net component. These projects are not counted as free-standing projects (i.e. they have a project count of 0) since the parent projects' objectives did not change, but their proportional commitments on social safety nets are considered in the portfolio analysis.

### NUMBER OF PROJECTS

The number of SSN projects almost doubled in the post-crisis period (Table 5.C.2). Projects managed under the oversight of the Social Protection sector board contributed around 60 percent of this increase, while SSN projects under other sector boards rose moderately.

## COMMITMENTS

The following criteria were used to calculate the SSN commitments in regular SSN operations:

- For projects with theme code 54, the SSN amount corresponds to the share of the theme code multiplied by the full dollar value of the project.

- For projects that had not been assigned code 54, IEG reviewed the design documents.

  - When the SSN was a component in the project and the project stated the dollar value per component, the amount of the SSN was taken as the total SSN commitment.

  - When the SSN was a subcomponent of the project or the dollar amount was not stated in the project design documents, the share assigned to the other social protection codes (51, 56, and 87) was considered the SSN share.

Most of the 33 Global Food Crisis Response (GFRP) operations with social safety net component(s) had been assigned entirely (100%) to the GFRP theme code (91). SSN commitments were calculated as follows:

- If a project had theme code 54, the SSN commitments were the percentage of total commitments assigned to that code.

- If a project had sector code JB (Other Social Services), the SSN commitments were the percentage of total commitments assigned to that sector code.

- If a project did not have a theme or sector code related to social protection, a share of the total commitments were assigned to SSNs manually.

Table 5.C.3 shows the distribution of SSN commitments by approval year and sector board. Changes in SSN commitments are much more pronounced than in number of projects. Total SSN commitments jumped from 1.4 billion in the pre-crisis period to 9.7 billion in the post-crisis period, a nearly six-fold increase. Consistent with the increasing pattern of the number of projects, commitments managed under the Social Protection sector rose dramatically from US$0.7 billion to nearly US$8 billion, contributing nearly 90 percent of the overall increase; while commitments under Other sectors experienced moderate increase.

A social safety net project may use any of six instruments, which were coded: conditional cash transfer; unconditional cash transfer; public works program; in-kind transfers; health and education subsidies; and energy, water, and other subsidies. These instruments are not mutually exclusive. Although nearly 60 percent of projects have only one instrument, around 35 percent of the projects adopted more than one social safety net instruments.

In addition, project design documents were reviewed and, for operations that were closed, Implementation Completion and Results Reports (ICRs) to code variables not available in the Bank's internal database, such as institutional development in social safety net operations.

## Social Safety Net Analytical and Advisory Activities (AAA)

For the AAA portfolio review, all of the economic and sector work (ESW) and non-lending technical assistance projects approved between FY2006 and FY2011 that had been assigned theme code 54 (Social Safety Nets) or PREM codes[5] managed under the oversight of Social Protection Sector were downloaded from the Bank's internal database (as of May 14, 2012). AAA products with thematic code 54 are regarded as Social Safety Nets products, those with PREM codes and managed under Social Protection sector board are classified as poverty reduction products.

A total of 289 activities were identified, 57 percent were ESW activities and the remaining 43 percent were non-lending technical assistance. Although ESW activities account for nearly 60 percent of all SSN AAA activities, their share declined dramatically from a peak of 87 percent in FY2007 to 43 percent in FY2011. Correspondingly, non-lending technical assistance activities experienced a significant increase from as low as 5 projects (13 percent) in FY2007 to 36 projects in both FY2010 and FY2011 (55–57 percent). Moreover, non-lending technical assistance activities exceeded ESW activities from FY2010 and the gap between these two widened in FY2011 (Table 5.C.4).

As of May 14, 2012, 19 percent of the ESW activities and 30 percent of the non-lending technical assistance activities were active.

Unlike lending portfolio, the majority of SSN AAA activities are managed by the Social Protection sector,[6] on average 71 percent. Although the number of SSN AAA activities managed by Other sectors is always smaller than that managed by the Social Protection sector, the increase in the share managed by Other sectors in the post-crisis period is more significant (77 versus 24 percent), which leads to a larger share of the overall SSN AAA activities (on average 32 percent in the post-crisis period compared to 25 percent in the pre-crisis period).

For both ESW and non-lending technical assistance activities, the dollar amounts correspond to expenditures and not commitments. Instead of looking into the overall expenditure of each activity, the review focused on the expenditures on Social Safety Nets and Poverty.[7] In order to conduct this specific analysis, the SSN (or Poverty) expenditure amount was calculated as the share of the theme code 54 (or PREM codes) multiplied by the full dollar expenditure of the activity. Adding these two gives the overall delivered amount related to the analysis.[8]

Total expenditure on SSN AAA activities increased moderately in the post-crisis period (51 percent), with similar magnitudes managed by Social Protection and Other sectors. However, changes in ESWs and non-lending TAs are totally different. Among activities managed by the Social Protection sector, expenditures on ESW decreased by 21 percent in FY2009–11, while non-lending TA activities received almost five times more expenditure, compared to FY2006–08. In contrast, expenditures on non-lending TA managed by Other sectors almost stagnated, whereas, that on ESW more than doubled (Table 5.C.6).

Beyond these activities approved in FY2006–11, a total of 79 AAA activities were approved in FY2012; 29 (37 percent) have been delivered to clients and 50 (63 percent) had not yet been delivered.[9] Among these, 31 were ESW activities and 48 were non-lending technical assistance. The delivery rate among ESW and non-lending technical assistance activities is consistent with the overall distribution. In addition, 43 activities (28 ESW and 15 TA) are in the pipeline for FY2013, and another five (four ESW and one TA activity) for FY2014.

## Endnotes

[1] IEG (Independent Evaluation Group). 2011. *Social Safety Nets: An Evaluation of World Bank Support, 2000–2010.* Washington, DC: Independent Evaluation Group, the World Bank Group.

[2] IEG evaluation on Social Safety Nets (2011) does not include recipient-executed, special financing, or supplemental financing projects.

[3] IEG evaluation on Social Safety Nets (2011) already included 7 global food crisis response projects approved in FY09 and FY10. Thus, total number of GFRP projects with SSN component is 33 in this portfolio.

[4] These 3 projects are: (1) Additional Financing for Colombia Social Safety Net Project (P104507): this project is a specific investment loan; managed under the Social Protection sector board. Its parent project (P089443) is included in both the current SSN portfolio and the previous SSN evaluation database. The main objective of the project is to consolidate and expand the Familias en Accion Conditional Cash Transfer program, and to improve M&E of the SSN program. (2) Sierra Leone Decentralized Service Delivery Program (P113757): this project is an adaptable program loan managed under the Social Protection sector board and financed by IDA. It focused on decentralized delivery of basic services, covering water, sanitation, education, health, etc. (3) Dominican Republic Additional Financing Social Sectors Investment Program (P116369): this project is a specific investment loan managed under the Social Protection sector board. The main objective is to mitigate the negative impact of the food/financial crisis on the poor and vulnerable. There are three main investment components: first, improve the Safety net response of the government by increasing coverage, effectiveness and transparency of the CCT

program and financing non-infrastructure supply equipments as needed; second, improve the scale and quality of the labor-based emergency response to the crisis; third, build a system of social monitoring of the current crisis impact and an early warning system for future crises.

[5] PREM codes refer to the following theme codes: 20 (Analysis of Economic Growth), 21 (Debt Management and Fiscal Sustainability), 22 (Economic Statistics, Modeling and Forecasting), 23 (Macroeconomic Management), 24 (Other Economic Management), 53 (Poverty Strategy, Analysis and Monitoring), and 59 (Gender).

[6] In the SSN lending portfolio, 45% of projects were managed under Social Protection sector board, compared to 55% under other sector boards. In terms of commitments, Social Protection sector board managed a little less amount than other sector boards in period FY2006–08, but much more (nearly five-fold) in FY2009–11.

[7] Many AAA pieces coded as poverty-reduction activities, in fact, include SSN activities, analyses, or components.

[8] There are 10 projects with missing information on the overall project expenditure amount. However, 7 of these 10 have separate information on expenditure amounts from Bank Budget and Trust Fund, which enables a re-calculation of the corresponding SSN/Poverty amounts, and thus fill the missing overall delivered amounts.

[9] IEG used the same selection criteria to identify SSN AAA activities in FY2012–14.

TABLE H.1 Number of Approved Projects with SSN Components, by Source

| Summary Table | Total |
|---|---|
| IEG Social Safety Net Evaluation (approved FY2006–10) | 130 |
| SSN projects approved in FY2011 | 34 |
| GFRP Projects with a SSN component | 26 |
| Special Funding Projects with a SSN theme code | 7 |
| Additional projects with the SSN theme code, approved in FY2006–10 | 3 |
| **TOTAL** | **200** |
| **SUPPLEMENTAL PROJECTS** | **23** |

SOURCE: World Bank Internal database.

TABLE H.2 Distribution of Projects with SSN Components by Approval Year (Number of Projects)

| Approval Year | Managed by the Social Protection Sector | Managed by Other Sectors | Total |
|---|---|---|---|
| 2006 | 8 | 12 | 20 |
| 2007 | 4 | 17 | 21 |
| 2008 | 14 | 13 | 27 |
| 2009 | 22 | 24 | 46 |
| 2010 | 25 | 27 | 52 |
| 2011 | 18 | 16 | 34 |
| 2006–08 | 26 | 42 | 68 |
| 2009–11 | 65 | 67 | 132 |
| **PERCENTAGE CHANGE** | 150 | 60 | 94 |

TABLE H.3 SSN Commitments by Approval Year (US$ Millions)

| Approval Year | Social Protection Sector | Other Sectors | Total |
|---|---|---|---|
| 2006 | 317.7 | 268 | 585.7 |
| 2007 | 218.2 | 288.2 | 506.4 |
| 2008 | 136.9 | 185.8 | 322.7 |
| 2009 | 3,080.7 | 575.1 | 3,655.8 |
| 2010 | 1,786.1 | 749.1 | 2,535.2 |
| 2011 | 3,123.4 | 398.3 | 3,521.7 |
| 2006–08 | 672.8 | 742 | 1,414.8 |
| 2009–11 | 7,990.2 | 1,722.5 | 9,712.7 |
| **PERCENTAGE CHANGE IN COMMITMENTS PRE- AND POST-CRISIS PERIODS** | 1,088 | 132 | 587 |
| Average Project Size 2006–08 | 25.9 | 17.7 | 20.8 |
| Average Project Size 2009–11 | 122.9 | 25.7 | 73.6 |
| **PERCENTAGE CHANGE IN PROJECT SIZE, PRE- AND POST-CRISIS** | 375 | 46 | 254 |

TABLE H.4 Regular and GFRP SSN Lending by Country Income Level

| Country Income Level | SSN Regular FY2006–08 | | SSN Regular FY2009–11 | | GFRP SSN FY2009–11 | |
|---|---|---|---|---|---|---|
| | Number Ops | Amount (US$ Millions) | Number Ops | Amount (US$ Millions) | Number Ops | Amount (US$ Millions) |
| HIC | | | 2 | 253.6 | | |
| LIC | 17 | 297.5 | 21 | 776.5 | 23 | 297 |
| LMIC | 23 | 543 | 44 | 1,724.4 | 10 | 226 |
| UMIC | 21 | 535.3 | 39 | 6,474.3 | | |
| TOTAL | 61 | 1,375.8 | 106 | 9,228.8 | 33 | 523 |

TABLE H.5 GFRP SSN Operations and Commitments by Region

| Region | SSN Regular FY2009–11 | | GFRP SSN FY2009–11 | |
|---|---|---|---|---|
| | Number Ops | Amount (US$ Millions) | Number Ops | Amount (US$ Millions) |
| AFR | 24 | 803.9 | 18 | 152.1 |
| EAP | 9 | 704.7 | 3 | 182.5 |
| ECA | 26 | 1,903.20 | 3 | 17 |
| LCR | 28 | 5,112.30 | 2 | 14 |
| MNA | 9 | 152.7 | 4 | 23.4 |
| SAR | 10 | 552 | 3 | 133.9 |
| TOTAL | 106 | 9,228.80 | 33 | 522.9 |

NOTE: AFR = Africa, EAP = East Asia and the Pacific, ECA = Europe and Central Africa, GFRP = Global Food Response Program, LCR = Latin America and the Caribbean, MNA = Middle East and North Africa, SAR = South Asia, SSN = social safety net.

SSN Instruments Post-Crisis: GFRP and Regular Portfolio

| Type of SSN Intervention/Instruments[a] | Number of Operations | |
|---|---|---|
| | Regular SSN FY2009–11 | GFRP SSN FY2009–11 |
| **Conditional Cash Transfer** | | |
| **Unconditional Cash Transfer** | 52 | 6 |
| **Public Works** | 32 | 16 |
| **In-Kind Transfers** | 11 | 17 |
| **Health and Education Subsidies** | 20 | 0 |
| **Water, Energy and Other Subsidies** | 16 | 0 |
| **TOTAL NUMBER OF PROJECTS** | 106 | 33 |

NOTE: a. These instruments are not mutually exclusive and therefore do not add to total number of projects.

TABLE H.7 Regular SSN Commitments by Lending Instrument and Country Income Level Pre- and Post-Crisis (US$ Millions)

| Lending Instrument | LICs | | MICs | | Total | |
|---|---|---|---|---|---|---|
| | Pre-Crisis | Post-Crisis | Pre-Crisis | Post-Crisis | Pre-Crisis | Post-Crisis |
| **DPO** | 71 | 14 | 364 | 2,194.0 | 435 | 2,208.0 |
| **IL** | 226.5 | 762.5 | 714.3 | 6,004.7 | 940.8 | 6,767.2 |
| **TOTAL** | 297.5 | 776.5 | 1,078.3 | 8,198.7 | 1,375.8 | 8,975.2 |

NOTE: This table excludes 2 DPLs in HICs in the post-crisis period ($253.6 million).
DPO = development policy operations, IL = investment lending, LICs = low-income countries, MICs = middle-income countries, SSN = social safety net.

**TABLE H.8** SSN Instruments Pre- and Post-Crisis Periods by Country Income Level

| Type of SSN Intervention/SSN Instruments[a] | Number of Operations with Instrument | | | | | |
| --- | --- | --- | --- | --- | --- | --- |
| | LICs | | | MICs | | |
| | FY2006–08 | FY2009–11 | Percentage Change | FY2006–08 | FY2009–11 | Percentage Change |
| Conditional Cash Transfer (CCT) | 2 | 1 | −50 | 14 | 30 | 114 |
| Unconditional Cash Transfer (UCT) | 3 | 9 | 200 | 10 | 41 | 310 |
| Public Works Program (PWP) | 3 | 12 | 300 | 6 | 20 | 233 |
| In-Kind Transfer | 4 | 3 | −25 | 7 | 8 | 14 |
| Health and Education Subsidies | 7 | 2 | −71 | 7 | 18 | 157 |
| Water, Energy and Other Subsidies | 1 | 2 | 100 | 9 | 14 | 56 |
| **TOTAL NUMBER OF PROJECTS** | 17 | 21 | | 44 | 83 | |

NOTE: LICs = low-income countries, MICs = middle-income countries, SSN = social safety net.
a. These instruments are not mutually exclusive and therefore do not add to total number of projects.

TABLE H.9 Institutional Development (ID) in Regular SSN Operations Pre- and Post-Crisis by Country Income Level

| Period | LICs | | | MICs | | | All | |
|---|---|---|---|---|---|---|---|---|
| | Number of Projects with ID | Total Number of Projects | Percentage of Projects with ID | Number of Projects with ID | Total Number of Projects | Percentage of Projects with ID | Total Number of Projects | Percentage of Projects with ID |
| FY06–08 | 8 | 17 | 47 | 26 | 44 | 59 | 61 | 56 |
| FY09–11 | 9 | 21 | 43 | 56 | 83 | 67 | 104 | 63 |
| **TOTAL** | **17** | **38** | **45** | **82** | **127** | **65** | **165ª** | **60** |

NOTE: LICs = low-income countries, MICs = middle-income countries.
a. Operations in HICs (2) are not included in this table so the total number of projects is 167.

TABLE H.10 GFRP SSN Lending by Country Vulnerability to a Food Price Crisis, FY09–11

| Country Vulnerability Level | Number of Countries | Commitments (US$ Millions) |
|---|---|---|
| Most Vulnerable | 11 | 78.6 |
| Vulnerable | 12 | 421.9 |
| Less Vulnerable | 3 | 14 |
| Other/NA | 1 | 8.4 |
| **TOTAL** | **27** | **522.9** |

NOTE: Country vulnerability list based on 87 LICs and lower-middle-income countries (LMICs).

TABLE H.11 Selected Performance Indicators for GFRP Social Safety Net Projects

| Project ID[a] | Economy | Number of People Employed | Number of Children Benefiting from School Feeding | Number of Girls Benefiting from Taking Home Rations | Pregnant and Lactating Women Receiving Nutritional Supplements or Education |
|---|---|---|---|---|---|
| P111545 | Kenya | | | | |
| P112017 | Djibouti | | | | |
| P112084 | Liberia | 17,000 | | | |
| P112107 | Liberia | | An average of 59,608 beneficiaries during the school year over the 3-year project period | An average of 2,894 beneficiaries a year | An average of 3,094 pregnant and lactating women a year |
| P112133 | Haiti | | | | |
| P112136 | Tajikistan | | | | 80,000 |
| P112142 | Kyrgyz Republic | | | | |
| P112345 | Republic of Yemen | | | | |
| P112761 | Bangladesh | | | | |
| P112908 | Moldova | | | | |

| Number of Households/ Individuals Benefiting from Cash Transfer Programs | Comments |
|---|---|
| 250,470 Individuals | |
| | The SSN component of this project focuses on identification of the poorest, targeting at least 5,000 poorest households. |
| | 17,000 people employed are the target. |
| | Target levels for number of children benefiting from school feeding, girls taking home rations, and women receiving nutritional supplements are 62,000, 4,300 a year, and 3,300 a year. |
| | No implementation status report available in the system. |
| | 80,000 women (50% of 160,000 targeted women) are targeted to receive nutrition education and practice exclusive breastfeeding for first 6 months. |
| | No implementation status report available in the system. |
| | No implementation status report available in the system. |
| | This project didn't measure number of people employed, but budget increased to food-related safety net programs. |
| 72,000 Households | 50,000 households benefiting from cash transfer program is the target. |

| Project ID° | Economy | Number of People Employed | Number of Children Benefiting from School Feeding | Number of Girls Benefiting from Taking Home Rations | Pregnant and Lactating Women Receiving Nutritional Supplements or Education |
|---|---|---|---|---|---|
| P113002 | Nepal | Public works benefited 180,758 households in FY08/09, 226,028 in 09/10, and 142,434 in 10/11. | | | |
| P113117 | West Bank and Gaza | | | | |
| P113134 | Madagascar | 305,079 | | | |
| P113141 | Sierra Leone | 35,785 | | | |
| P113219 | Sierra Leone | | | | |
| P113221 | Central African Republic | | 127,316 | | |
| P113224 | Madagascar | | | | |
| P113438 | Burundi | | | | |

| Number of Households/ Individuals Benefiting from Cash Transfer Programs | Comments |
|---|---|
| | Numbers of households benefited from the public works programs are combined results with the additional financing project (P120538). |
| 64,000 Households | There is no target set for number of households benefitting from cash transfer program. |
| | Target level for number of people employed is 160,000. |
| | Target level for number of people employed is 31,000. |
| | 8,200 lactating mothers and children under five, in hospitals administered by the Ministry of Health and Sanitation that receive feedings three times a day; 3,470 pupils in government boarding schools and handicapped children are fed three times per day; 380 children in remand homes and approved schools received food. |
| | Target level for number of children benefitting from school feeding is 153,000. |
| | No SSN indicators in the ICR for parent project (P105135). |
| | Prior to the Food Crisis Response DPG, there was no allocation in the national budget aimed at the School Feeding Program. As of July 31, 2009, the 2009 budget allocated about $5 million to this program. Prior to Food Crisis Response DPG, the WFP school meal program was implemented in 269 primary schools, 238,873 children were covered. As of July 31, 2009, feeding was implemented in additional 60 schools, 120,000 children benefited from additional hot meals distributed in schools. |

| Project IDª | Economy | Number of People Employed | Number of Children Benefiting from School Feeding | Number of Girls Benefiting from Taking Home Rations | Pregnant and Lactating Women Receiving Nutritional Supplements or Education |
|---|---|---|---|---|---|
| P113468 | Guinea-Bissau | | | | |
| P113492 | Philippines | | | | |
| P113544 | Togo | | | | |
| P113586 | South Sudan | 25,303 | | | |
| P113608 | Guinea | 227,000 | | | |
| P114441 | Nicaragua | | 558,365 Primary School Children | | |
| P114683 | Ethiopia | | | | |
| P114740 | Ethiopia | 4,343 | | | |
| P114863 | Lao PDR | | | | |

| Number of Households/ Individuals Benefiting from Cash Transfer Programs | Comments |
|---|---|
| | Target 14,000 students receiving one meal a day and 160,000 work days in food for work activities achieved, current levels are 13,812 students and 285,000 work days. |
| Over 1 Million Families | The goal of reaching 320,000 poor households by the end of 2008 was attained. |
| | A total of 40,458 school children in 178 schools currently benefit from school feeding sub-component, financed by P113544 and another additional financing grants (US$8.7 million). |
| | Target level for number of participants in public works program is 25,462. |
| | Target level is 5,300 temporary jobs created for a period of 2–3 months (300,000 beneficiaries); 7 billion GNP (1M US$) distributed in wages to workers in the program. |
| | Target level for primary school children receiving lunches in targeted areas is 216,627. In addition, 50,777 pre-school children also receiving lunches in targeted areas compared to the target 35,411 children. Overall, number of days children received school lunches reached 172, higher than the target 133 days. |
| | Productive Safety Net Program (PSNP) reached 7.2 million beneficiaries in 2007, 7.4 million in 2008 and 7.6 million in 2009. 83% of beneficiaries participated in public works and 17% benefitted from direct support. PSNP financed 34,000 public works projects annually. |
| | Target level for number of people employed is 3,500. In addition, this project created 108,425 person-days of employment, higher than the target 90,000 person-days. |
| | No related indicators after project restructuring. |

| Project ID[a] | Economy | Number of People Employed | Number of Children Benefiting from School Feeding | Number of Girls Benefiting from Taking Home Rations | Pregnant and Lactating Women Receiving Nutritional Supplements or Education |
|---|---|---|---|---|---|
| P115938 | Lao PDR | | | | |
| P115952 | Tanzania | | | | |
| P117203 | Cambodia | | | | |
| P117320 | West Bank Gaza | | | | |
| P118226 | Sierra Leone | 16,110 | | | |
| P120538 | Nepal | | | | |

NOTE: a. Project names are in Appendix Table 3.B.
DPG = development policy grant, GFRP = Global Food Response Program, GNP = gross national product,
ISR = implementation status report, SSN = social safety net, WFP = World Food Programme.

| Number of Households/ Individuals Benefiting from Cash Transfer Programs | Comments |
|---|---|
| 55,323 beneficiaries (all female) of cash transfer program | Target level for cash transfer beneficiaries is 50,000. In addition, 95% of targeted children in primary education received weekly micronutrient supplements and deworming medication twice a year (above the target 80%). |
| | No ISR available in system. |
| | Emergency distribution of rice to 342,853 people in 200 communes (higher than target 300,000). No other indicators. |
| | No separate results reported in parent project ISR (P081477). |
| | Parent project (P079335) created temporary employment for 16,515 people, plus additional 3,160 beneficiaries of the Cash for Work pilot program. Overall, the entire project created temporary employment for 35,785 people, exceeding the target level 31,000 people. |
| | Number of households benefited from the public works programs is combined with the results from parent project (P113002). |

**TABLE H.12** Social Protection Portfolio Performance FY2006–11

| Rating, Instrument, and Managing Sector | | FY2006–08 | | FY2009–11 | |
|---|---|---|---|---|---|
| | | Number of Projects Rated | Percentage Rated Satisfactory[a] | Number of Projects Rated | Percentage Rated Satisfactory[a] |
| Development Outcome, Overall Portfolio (DPOs + ILs) | Social Protection | 41 | 83 | 25 | 72 |
| | Human Development | 189 | 68 | 131 | 70 |
| | Bank-wide | 741 | 79 | 506 | 72 |
| ILs Only—Development Outcome | Social Protection | 34 | 79 | 21 | 76 |
| | Human Development | 169 | 66 | 125 | 70 |
| | Bank-wide | 588 | 78 | 424 | 70 |
| ILs Only—Bank Quality at Entry | Social Protection | 34 | 71 | 21 | 62 |
| | Human Development | 169 | 67 | 125 | 58 |
| | Bank-wide | 588 | 73 | 424 | 61 |
| ILs Only—Bank Quality of Supervision | Social Protection | 34 | 79 | 21 | 76 |
| | Human Development | 169 | 75 | 125 | 77 |
| | Bank-wide | 588 | 85 | 424 | 78 |

| Rating, Instrument, and Managing Sector | | FY2006–08 | | FY2009–11 | |
|---|---|---|---|---|---|
| | | Number of Projects Rated | Percentage Rated Satisfactory[a] | Number of Projects Rated | Percentage Rated Satisfactory[a] |
| DPOs Only—Development Outcome | Social Protection | 7 | 100 | 4 | 50 |
| | Human Development | 20 | 80 | 6 | 67 |
| | Bank-wide | 152 | 80 | 82 | 83 |
| DPOs Only—Bank Quality at Entry | Social Protection | 7 | 63 | 4 | 75 |
| | Human Development | 20 | 81 | 6 | 83 |
| | Bank-wide | 152 | 84 | 82 | 89 |
| DPOs Only—Bank Quality of Supervision | Social Protection | 7 | 100 | 4 | 50 |
| | Human Development | 20 | 90 | 6 | 67 |
| | Bank-wide | 152 | 92 | 82 | 89 |

NOTE: IL=Investment Lending, DPO=Development Policy Operation.
a. Percent rated moderately satisfactory or higher.

Distribution of SSN Commitments among SSN GFRP Projects by Economy

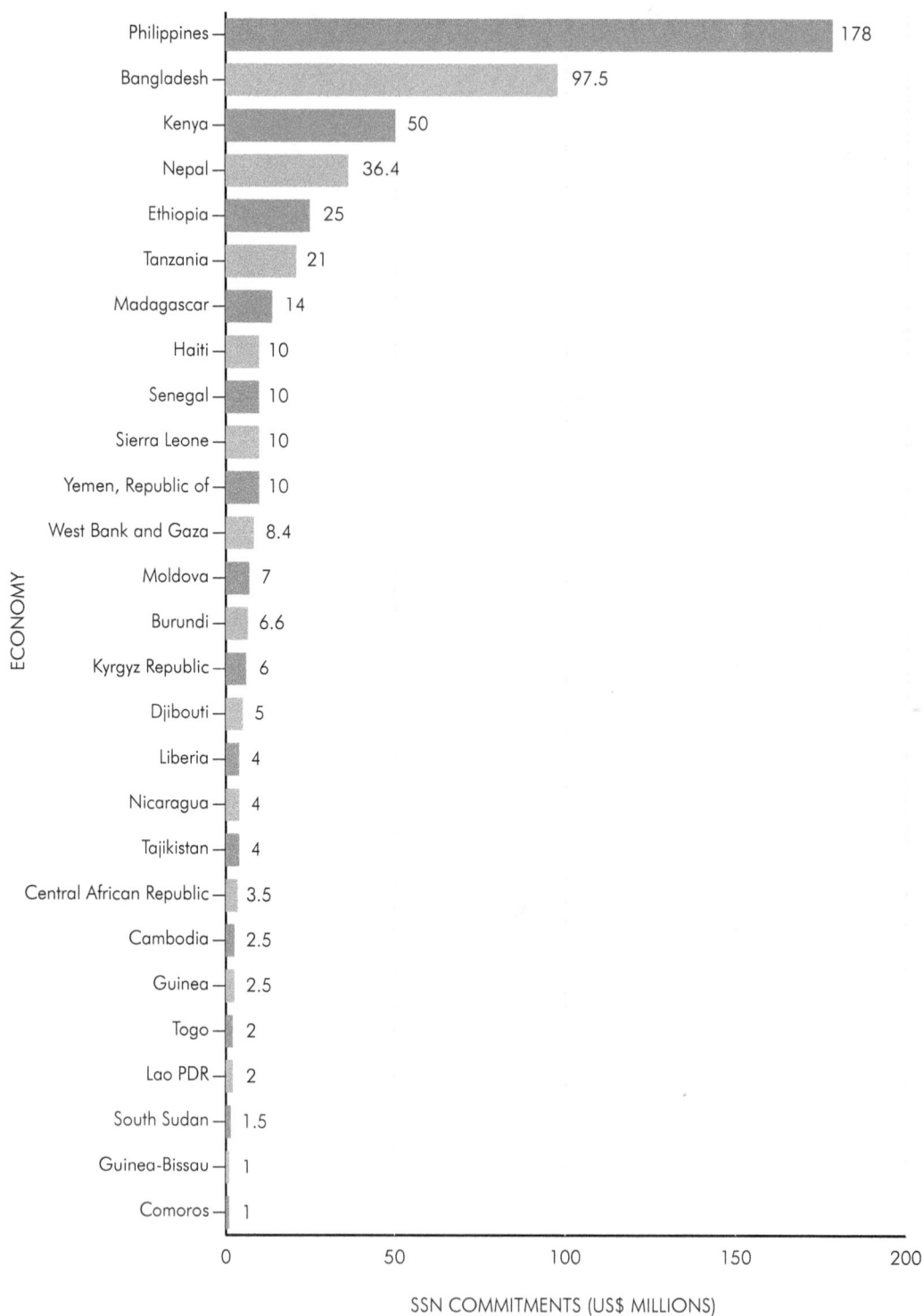

| Economy | SSN Commitments (US$ millions) |
|---|---|
| Philippines | 178 |
| Bangladesh | 97.5 |
| Kenya | 50 |
| Nepal | 36.4 |
| Ethiopia | 25 |
| Tanzania | 21 |
| Madagascar | 14 |
| Haiti | 10 |
| Senegal | 10 |
| Sierra Leone | 10 |
| Yemen, Republic of | 10 |
| West Bank and Gaza | 8.4 |
| Moldova | 7 |
| Burundi | 6.6 |
| Kyrgyz Republic | 6 |
| Djibouti | 5 |
| Liberia | 4 |
| Nicaragua | 4 |
| Tajikistan | 4 |
| Central African Republic | 3.5 |
| Cambodia | 2.5 |
| Guinea | 2.5 |
| Togo | 2 |
| Lao PDR | 2 |
| South Sudan | 1.5 |
| Guinea-Bissau | 1 |
| Comoros | 1 |

**TABLE H.13** SSN Analytical and Advisory Activities by Approval Year and Product Line

| Approval Year | ESW | | Non-Lending TA | | Total |
|---|---|---|---|---|---|
| | Number of Activities | Percentage of Total Activities | Number of Activities | Percentage of Total Activities | |
| FY2006 | 26 | 79 | 7 | 21 | 33 |
| FY2007 | 34 | 87 | 5 | 13 | 39 |
| FY2008 | 26 | 52 | 24 | 48 | 50 |
| FY2009 | 21 | 55 | 17 | 45 | 38 |
| FY2010 | 30 | 45 | 36 | 55 | 66 |
| FY2011 | 27 | 43 | 36 | 57 | 63 |
| **TOTAL** | 164 | 57 | 125 | 43 | 289 |

NOTE: ESW = economic and sector work, TA = technical assistance.

TABLE H.14 SSN Analytical and Advisory Activities by Approval Year and Sector Board

| Approval Year | Managed by the SP Sector | | Managed by Other Sectors | | Total |
|---|---|---|---|---|---|
| | Number of Activities | Percentage of Total Activities | Number of Activities | Percentage of Total Activities | |
| FY06 | 23 | 70 | 10 | 30 | 33 |
| FY07 | 33 | 85 | 6 | 15 | 39 |
| FY08 | 36 | 72 | 14 | 28 | 50 |
| FY09 | 25 | 66 | 13 | 34 | 38 |
| FY10 | 41 | 62 | 25 | 38 | 66 |
| FY11 | 48 | 76 | 15 | 24 | 63 |
| **TOTAL** | 206 | 71 | 83 | 29 | 289 |
| FY2006–08 | 92 | 75 | 30 | 25 | 122 |
| FY2009–11 | 114 | 68 | 53 | 32 | 167 |
| **PERCENTAGE CHANGE** | 24 | -9 | 77 | 29 | 37 |

NOTE: FY = fiscal year, SP = social protection, SSN = social safety net.

TABLE H.15 Distribution of SSN Analytical and Advisory Activities Expenditure (US$ Millions)

| Approval Year | ESW | | | Non-Lending TA | | | Total Expenditure | | |
|---|---|---|---|---|---|---|---|---|---|
| | SP Sector | Other Sectors | Total | SP Sector | Other Sectors | Total | SP Sector | Other Sectors | Total |
| FY2006 | 1.5 | 0.8 | 2.3 | 0.1 | 0.1 | 0.2 | 1.6 | 0.9 | 2.5 |
| FY2007 | 2.5 | 0.1 | 2.6 | 0.1 | 0.1 | 0.2 | 2.6 | 0.2 | 2.9 |
| FY2008 | 2.5 | 0.3 | 2.8 | 0.8 | 1.4 | 2.2 | 3.3 | 1.7 | 5 |
| FY2009 | 0.6 | 0.8 | 1.3 | 0.8 | 0.2 | 1 | 1.4 | 0.9 | 2.3 |
| FY2010 | 1.9 | 1.4 | 3.3 | 2.2 | 1 | 3.2 | 4.1 | 2.4 | 6.5 |
| FY2011 | 2.6 | 0.7 | 3.3 | 3.1 | 0.5 | 3.6 | 5.7 | 1.2 | 6.9 |
| FY2006–08 | 6.5 | 1.3 | 7.8 | 1 | 1.6 | 2.6 | 7.5 | 2.9 | 10.4 |
| FY2009–11 | 5.1 | 2.9 | 8 | 6.1 | 1.7 | 7.7 | 11.2 | 4.5 | 15.7 |
| PERCENTAGE CHANGE | -21 | 123 | 3 | 479 | 5 | 195 | 48 | 58 | 51 |

NOTE: ESW = economic and sector work, FY = fiscal year, SP = social protection, SSN = social safety net, TA = technical assistance.

# Appendix I
# Summary Results of Country Case Studies

TABLE I.1  Country Case Studies Consolidated Summary

1. How did price developments in [country] during 2007–08 relate to the international trend?

| Answer | Frequency | Percent |
|---|---|---|
| Higher | 5 | 25 |
| Lower | 12 | 60 |
| N/A | 1 | 5 |
| Same | 2 | 10 |
| TOTAL | 20 | 100 |

2. Social risk management instruments in country (percent, n=20)

| Instruments | Unconditional Cash Transfer | Conditional Cash Transfer | In-Kind[a] | Targeted subsidies | Public Works Program | NHP | Contributory Programs | LMP | General Subsidies |
|---|---|---|---|---|---|---|---|---|---|
| Percent | 60 | 50 | 80 | 65 | 75 | 65 | 50 | 30 | 55 |

NOTE: Answers are not mutually exclusive. a. Includes school feeding.

| Question | Answer (Percent), n=20 | | | |
| --- | --- | --- | --- | --- |
| | Yes | No | N/A | Total |
| 3. Did the food crisis have a measurable impact on poverty? | 85 | 10 | 5 | 100 |
| 4. Did the government enact immediate food price reducing policies such as elimination of tariff/taxes? | 75 | 20 | 5 | 100 |
| 5. Did the government enact limitations on food exports? | 45 | 35 | 20 | 100 |
| 6. Did the country enact measures for immediate (next season) agricultural supply response through expansion of input distribution/subsidies (for fertilizers, seeds) or other output subsidies to agricultural food producers? | 90 | 10 | 0 | 100 |
| 7. Did the government reallocate fiscal resources? | 65 | 10 | 15 | 100 |

8. Did the Bank conduct a rapid response diagnosis of the social protection system or an assessment of the potential for agricultural supply response?

| Answer | SP | Percent | ARD | Percent |
| --- | --- | --- | --- | --- |
| Yes | 2 | 10 | 10 | 50 |
| No | 13 | 65 | 8 | 40 |
| No But Other Donors Did | 2 | 10 | 0 | 0 |
| No But There Was Previous AAA | 3 | 15 | 2 | 10 |
| TOTAL | 20 | 100 | 20 | 100 |

9. Assess the relevance of SSN and ARD recommendations for the immediate future and the longer term (out the ones with diagnosis)

| Answer | Relevance | | | | Effectiveness | | | |
|---|---|---|---|---|---|---|---|---|
| | SP | Percent | ARD | Percent | SP | Percent | ARD | Percent |
| Very Relevant or Effective | 5 | 25 | 5 | 25 | 1 | 6 | 1 | 5 |
| Relevant or Effective | 0 | 0 | 5 | 25 | 8 | 41 | 3 | 15 |
| Partially Relevant or Effective | 1 | 5 | 2 | 10 | 3 | 12 | 4 | 20 |
| Not at all Relevant or Effective | 0 | 0 | 1 | 5 | 0 | 0 | 5 | 25 |
| No Diagnosis | 14 | 70 | 7 | 35 | 8 | 41 | 7 | 35 |
| TOTAL | 20 | 100 | 20 | 100 | 20 | 100 | 20 | 100 |

10. Did the Bank introduce new lending and non-lending?
   Did it adjust on-going activities or proceed with business as usual?

| Answer | Frequency | Percent |
|---|---|---|
| Introduced New Lending | 19 | 95 |
| Introduced New Non-Lending | 13 | 65 |
| Adjusted On-Going Activities | 16 | 80 |
| Proceed with Business as Usual | 2 | 10 |

11. To what extent, did the bank provide policy advice on immediate and long term response to the crisis in the social protection and agricultural sector?

| | Answer (Percent), n=20 | | | | Total |
|---|---|---|---|---|---|
| | To a Great Extent | Somewhat | A Little | Not At All | |
| Immediate Response | 30 | 50 | 15 | 5 | 100 |
| Long-Term Response | 40 | 45 | 10 | 5 | 100 |

12. To what extent was the support to the 2007–08 global food crisis in [country] coordinated with other donors in GFRP activities and regular portfolio?

| | Answer (Percent), n=20 | | | | Total |
|---|---|---|---|---|---|
| | To a Great Extent | Somewhat | A Little | Not At All | |
| GFRP Activities | 35 | 55 | 10 | 0 | 0 |
| Regular Portfolio | 35 | 45 | 10 | 5 | 5 |

| Question | | Answer (Percent), n=20 | | | Total |
|---|---|---|---|---|---|
| | | Yes | Somewhat | No | |
| 13. Was there a trade-off between speed of preparation and quality of the intervention's design or its implementation challenges? | | 20 | 15 | 65 | 100 |
| 14. Was there a shift in regular WBG activities (both lending and non-lending) in [country] towards building greater resilience to future food crisis? | Social Protection | 85 | | 15 | 100 |
| | Agriculture and Rural Development | 40 | | 60 | |
| 15. Are Bank-supported activities in the rural and agricultural sector in the post crisis years more oriented towards increasing production capacity (in particular staple food production), storage, processing, and marketing? | | 55 | 10 | 35 | 100 |
| 16. Has the WBG's post-crisis policy dialogue with the government of [country] placed greater emphasis on building resilience to future crises, as evidence in CEMs and other modalities of interacting with decision makers? | | 70 | 5 | 25 | 100 |

# Bibliography

**ADB Independent Evaluation Department**. 2011. Special Evaluation Study on *Real-time Evaluation of Asian Development Bank's Response to the Global Economic Crisis of 2008–2009.*

**African Development Bank**, 2008. *The African Food Crisis Response.*

**Alam, Asad**. 2008. *"High food prices: Challenges and opportunities for ECA countries."* The World Bank.

**Alderman, Harold** and **Donald Bundy**. 2011. *"School feeding programs and development: are we framing the question correctly?"* The World Bank Research Observer.

**Aquila Communiqué**. 2009. "L'Aquila" Joint Statement on Global Food Security L'Aquila Food Security Initiative (AFSI).

**Bank Information Center**. 2008. *"Amid food riots and shaken governments IFIs scramble to develop a coherent response."*

**Berg, A., et al.**, 2010. *"Global Shocks and their Impact on Low- Income Countries: Lessons from the Global Financial Crisis."* Washington, D.C: IMF.

**Brahmbhatt, Milan** and **Luc Christiaensen**. 2008. *"Rising Food Prices in East Asia: Challenges and Policy Options"* World Bank Report No. 44998.

**Compton, Julia, Steve Wiggins**, and **Sharada Keats**. 2010. *"Impact of the global food crisis on the poor: what is the evidence?"* ODI. Table 1 p. 18 for specific findings in 6 country studies (Bangladesh, Cambodia, Guinea, Kenya, Lesotho, Swaziland) based on 2008–2009 surveys.

**Delgado, Christopher, Robert Townsend, Iride Ceccacci, Yurie Tanimichi Hoberg, Saswati Bora, Will Martin, Don Mitchell, Don Larson, Kym Anderson**, and **Hassan Zaman**. 2011. *"Food Security: The Need for Multilateral Action."* Postcrisis growth and development: a development agenda for the G-20. S. Fardoust, Y. Kim and C. Sepúlveda (eds.), Ch. 9 pp. 383–425.

**Dewees, Peter A., Bruce M. Campbell, Yemi Katerere, Almeida Sitoe, Anthony B. Cunningham, Arild Angelsen**, and **Sven Wunder**. 2011. *"Managing the Miombo Woodlands of Southern Africa: Policies, incentives and options for the rural poor."* Washington DC: Program on Forests (PROFOR).

**Didier, T., Hevia C.**, and **Schmukler, S**. 2010. *"How Resilient Were Developing Countries to the Global Crisis?"* Washington, D.C.: World Bank.

**FAO**. 2007. *Food Outlook Global Market Analysis.*

\_\_\_\_. 2008a. *FAO calls for urgent steps to protect the poor from soaring food prices.*

\_\_\_\_. 2008b. *High Cereal Prices Are Hurting Vulnerable Populations in Developing Countries.*

\_\_\_\_. 2008c. *Initiative on Soaring Food Prices.*

**FAO, IFAD, IMF, OECD, UNCTAD, WFP, WB, WTO, IFPRI, UN HLTF**. 2011. *Price Volatility in Food and Agricultural Markets: Policy responses.* Rome: FAO.

**Financial Times**. 2008. "Zoellick calls for fight against hunger to be global priority."

Fiszbein Ariel, Ringold Dena, Srinivasan Santhosh. 2011. *"Cash Transfers, Children and the Crisis: Protecting Current and Future Investments."* Washington, DC: World Bank.

G20 Leaders Summit—Final Communiqué. 2011.

Global Agriculture and Food Security Program Annual Report. 2011.

Grosh, Margaret, Colin Andrews, Rodrigo Quintana, Claudia Rodriguez-Alas. 2011. *Assessing Safety Nets Readiness in Response to Food Price Volatility.* The World Bank. Social Protection Discussion Paper 1118.

Grosh, M., C. del Ninno, E. Tesliuc, and A. Ouerghi. 2008. *"For Protection and Promotion, The Design and Implementation of Effective Safety Nets."* Washington, DC: World Bank.

Headey, D. and Shenggen Fan. 2008. *"Anatomy of a crisis: the causes and consequences of surging food prices."* Agricultural Economics, International Association of Agricultural Economists, 39(s1), 375–391.

IFC, Development Outcome Tracking System. 2012.

IEG. 2008. *Lessons from World Bank Group Responses to Past Financial Crises.* Evaluation Brief 6, Washington, DC.: World Bank.

____. 2009a. *Project Evaluation Methodology—Investment Operations in the Independent Evaluation of IFC's Development Results.* Washington, D.C.: World Bank.

____. 2009b. *Knowledge for Private Sector Development—Enhancing the Performance of IFC's advisory Services.* Washington, D.C.: World Bank.

____. 2009c. *The World Bank Group's Response to the Global Crisis—Update on an Ongoing IEG Evaluation.* Evaluation Brief 8. Washington, D.C., World Bank.

____. 2010. *The World Bank Group's Response to the Global Economic Crisis—Phase I.* Washington, D.C., World Bank.

____. 2011a. *Growth and Productivity in Agriculture and Agribusiness. Evaluative Lessons from the World Bank Experience.* Washington, D.C.: World Bank.

____. 2011b. *Social Safety Nets: An Evaluation of World Bank Support, 2000–2010.* Washington, D.C.: World Bank.

____. 2012. *Results and performance of the World Bank Group 2012.* Washington, D.C.: World Bank.

IMF. 2008. *IMF Closely Involved in Drive to Relieve Global Food Crisis.* IMF Survey Online.

IMF and the World Bank. 2009. *The Implications of the Global Financial Crisis for Low-Income Countries—An Update.* Washington, D.C.

____. 2010a. *How Resilient Have Developing Countries Been During the Global Crisis?* Washington, D.C.

____. 2010b. *How Did Emerging Markets Cope in the Crisis,* paper prepared by the Strategy, Policy, and Review Department. Washington, D.C.

____. 2010c. *Emerging from the Global Crisis: Macroeconomic Challenges Facing Low-Income Countries,* paper prepared by the Strategy, Policy, and Review Department. Washington, D.C.

Ivanic, Maros and Will Martin. 2008. *Implications of Higher Global Food Prices for Poverty in Low-Income Countries.* Policy Research Working Paper 4594. The World Bank, Washington.

Joint Multilateral Development Banks' Action Plan for Improving Coordination on Food and Water Security. October 2011.

Ministerial Declaration: Action Plan on Food Price Volatility and Agriculture Meeting of G20 Agriculture Ministers. Paris, 2011.

Mitchell, Donald. 2008. "A Note on Rising Food Prices" Policy Research Working Paper 4682. The World Bank, Washington.

Pan, Lei and Luc Christiaensen. 2011. "Who is Vouching for the Input Voucher? Decentralized Targeting and Elite Capture in Tanzania", Policy Research Working Paper 5651. The World Bank.

Sabates-Wheeler, Rachel, Stephen Devereux. 2010. "Cash Transfers and High Food Prices: Explaining Outcomes on Ethiopia's Productive Safety Net Programme." Future Agricultures Working Paper 004.

The Paris Declaration on Aid Effectiveness (2005) and the Accra Agenda for Action (2008).

Timmer, C. Peter. 2010. Reflections on food crises past Food Policy 35(1).

Tiwari, Sailesh, and Hassan Zaman. 2010. "The Impact of Economic Shocks on Global Undernourishment." The World Bank, Policy Research Working Paper 5215.

Von Braun, J. and Gebreyohanes G.T. 2012. Global Food Price Volatility and Spikes: An Overview of Costs, Causes, and Solutions. ZEF-Discussion Papers on Development Policy, Bonn.

Von Braun, J. and Torero, M. 2009. Exploring the Price Spike. Choices 24(1).

WFP and the World Bank. 2009. Rethinking School Feeding—Social Safety Nets, Child Development and the Education Sector.

World Bank. 2005. Managing Food Price Risks and Instability in an Environment of Market Liberalization. World Bank, Agriculture and Rural Development. Washington, D.C.

____. 2008a. Framework Document for Proposed Loans, Credits and Grants for a Global Food Crisis Response Program.

____. 2008b. Rising Food Prices. The World Bank's Latin America and Caribbean Region Position Paper. Report No. 44718.

____. 2008c. Rising Food and Fuel Prices: Addressing the Risks to Human Capital. Human Development Network (HDN), Poverty Reduction and Economic Management (PREM) Network.

____. 2008d. Guidance for Responses from the HD Sector to Rising Food and Fuel prices. Human Development Network (HDN).

____. 2008e. World Bank President Calls for Plan to Fight Hunger in Pre-Spring Meetings Address. World Bank News Release.

____. 2008f. Rising Food Prices Threaten Poverty Reduction. World Bank News Release.

____. 2008g. Rising Food Prices: Policy Options and World Bank Response.

____. 2008h. Benin—Emergency Food Security Support Project.

____. 2008i. Global Food Crisis Response Program.

____. 2008j. Togo—Community Development Project: additional financing.

____. 2008k. Addressing the Food Crisis: The Need for Rapid and Coordinated Action.

____. 2008l. Republic of Yemen, Emergency Additional Financing Grant for the Third Social Fund for Development. Report No. 44043.

____. 2008m. Mozambique: Fifth Poverty Reduction Support Credit. Report No. 44646.

____. 2008n. Proposed Credit and Grant to the Republic of Mozambique for a Fifth Poverty Reduction Support Credit. Report No. 44846.

____. 2009a. Rapid Social Response Monthly Report. June 2009.

_____. 2009b. *Implementing Agriculture for Development: World Bank Group Agriculture Action Plan FY2010–2012.*

_____. 2009c. *Kenya Economic Outlook.*

_____. 2009d. *Tanzania—Accelerated Food Security Program.*

_____. 2010a. *Food Price Increases in South Asia: National Responses and Regional Dimensions. South Asian Region.*

_____. 2010b. *Kenya Agricultural Input Supply Project: Implementation Completion an Results Report (ICR000148).*

_____. 2010c. *Senegal—Additional financing for Food Security (GFRP) Project: Emergency Project Paper.*

_____. 2010d. *Progress Report—Global Food Crisis Response Program.*

_____. 2011a. *Rising Food and Energy Prices in Europe and Central Asia.* Report No. 61097.

_____. 2011b. Harmonized List of Fragile Situations FY2012.

_____. 2011c. *World Development Report 2011.*

_____. 2011d. *Conflict, Security and Development.*

_____. 2011e. *Rising Food and Energy prices in Europe and Central Asia.* Report No. 61097.

_____. 2011f. *Tajikistan: Delivering Social Assistance to the Poorest Households.* Report No. 56593.

_____. 2011g. *Tajikistan: Social Safety Net Strengthening Project.* Report No. 61114.

_____. 2012a. The Global Food Crisis.

_____. 2012b. Food Price Watch.

_____. 2012c. *Rapid Social Response Program Progress Report 2012.*

_____. 2012d. *World Bank Group Agriculture Action Plan 2013–2015.*

_____. 2012e. Global Monitoring Report.

_____. 2012f. *Resilience, Equity and Opportunity The World Bank's Social Protection and Labor Strategy 2012–2022.* Consultations Report.

_____. 2012g. *Responding to higher and more volatile world food prices.* Report No. 68420.

_____. 2012h. *Project Performance Assessment Report—Burundi: Food Crisis Response Development Policy Grant.* Report No. 67471.

**World Bank. IFAD. FAO**. 2009. *Improving Food Security in Arab Countries.*

**World Food Program**. *Vouchers and Cash Transfers as Food Assistance Instruments: Opportunities and Challenges.* Rome, 2008.

**Wodon, Quentin** and **Hassan Zaman**. 2008 "*Rising Food Prices in Sub-Saharan Africa: Poverty Impact and Policy Responses*" Policy Research Working Paper 4738. The World Bank, Washington.

**Yemtsov, Ruslan**. 2008. "*The Food Crisis: Global Perspectives and Impact on MENA—Fiscal and Poverty Impact.*"

www.ingramcontent.com/pod-product-compliance
Lightning Source LLC
Chambersburg PA
CBHW080607270326
41928CB00016B/2958